Human Anatomy & Physiology

Study Notes

Adeyemi Olubummo

Human Anatomy and Physiology
Study Notes

iUniverse books may be ordered through booksellers or by contacting:

iUniverse
1663 Liberty Drive
Bloomington, IN 47403
www.iuniverse.com
1-800-Authors (1-800-288-4677)

ISBN: 978-1-4502-3552-5 (sc)
ISBN: 978-1-4502-3553-2 (ebook)

Printed in the United States of America

iUniverse rev. date: 7/29/2010

Dedication

To my wife, Catherine, for her patience, support, and encouragement.

About the Author

Adeyemi Olubummo has a B.S. in biology from Fairleigh Dickinson University and an M.S. from Arnold and Marie Schwartz College of Pharmacy of Long Island University. Prior to taking up college teaching, Adeyemi worked as a research scientist at Unilever and Continental Pharmaceuticals, focusing on product development and manufacturing of drugs and household products. His knowledge of pharmacology, pharmacokinetics, and drug design gives him an added advantage in understanding the interdependencies and interactions of various systems of the human body. He has taught anatomy and physiology at various campuses of the City University of New York since 1994.

Contents

Chapter 22: **The Respiratory System**

References

Preface

This study guide on anatomy and physiology is designed to accompany any major anatomy and physiology textbook typically used by nursing students. It is designed to facilitate students' quick reference to important concepts covered in lectures. Most of the important concepts in a typical textbook are covered using an easy-to-understand, bullet-point format. The book is not intended to replace a typical anatomy and physiology textbook but to serve as an adjunct to it. By using this book at home, the student will save countless hours that would otherwise be needed to comb through a traditional textbook of over 1,000 pages. The fill-in-the-blank questions are designed to check students' understanding of the materials covered in each chapter. Answers to these questions are provided at the end of the book.

Chapter 1

Introduction to the Human Body

Anatomy: Anatomy can be divided into gross (macroscopic) anatomy and microscopic anatomy

1.1 Branches of Anatomy and Physiology
1.1.1 Gross Anatomy

In gross anatomy we study relatively large structures, such as the heart, kidney, and lungs, without the aid of a microscope. There are many different branches of gross anatomy:

1. Surface anatomy is the study of general form and superficial markings.
2. Regional anatomy focuses on anatomical organization of specific areas of the body.
3. Systemic anatomy is the study of the structure of organ systems, such as the digestive system.
4. Developmental anatomy deals with the study of the changes in forms that occur between conception and physical maturity.
5. Clinical anatomy focuses on different areas of specialties in clinical practice—e.g., medical anatomy, radiographic anatomy, and surgical anatomy.

1.1.2 Microscopic Anatomy

1. Cytology is the study of the internal structure of individual cells with the aid of microscopes.
2. Histology is the branch of anatomy that deals with the study of tissues.

1.1.3 Physiology

The study of the functions of the human body

1. Cell physiology is the study of the functions of cells.
2. Special physiology is the study of the physiology of special organs.
3. Systemic physiology is the study of the functioning of specific organ systems.
4. Pathological physiology is the study of the effects of diseases on organs or organ systems' functions.

1.1.4 Levels of Organization

1. The chemical (or molecular) level: atoms and molecules
2. The cellular level: molecules interacting to form organelles
3. The tissue level: groups of cells working together to perform one or more specific functions
4. The organ level: two or more tissues working in combination to perform several functions
5. The organ system level: organs interacting in organ systems
6. The organism level: an organism, such as a human

1.1.5 Necessary Life Functions

1. **Maintaining Boundaries**: All the cells in the body are surrounded by selectively permeable membranes keeping unwanted materials out.
2. **Movement:** All activities in the body facilitated by the muscular system, such as blood

circulation or propelling food through the digestive system. Movement is carried out by entire organisms, organs, tissues, cells, or organelles.

3. **Responsiveness**: Ability to sense changes in the environment and respond to them accordingly. The organism detects stimulus, and the effectors respond to the stimulus.

4. **Metabolism**: It involves all the chemical reactions occurring in the body cells. It includes breaking down substances (catabolism) and synthesizing complex substances from simpler ones (anabolism). Metabolism depends on the digestive and respiratory systems to make oxygen and nutrients available and on the cardiovascular system, which transports them throughout the body.

5. **Growth:** Increase in the size of a body part or an organism. For true growth to occur in cells, constructive activities in the cells must supersede the destructive activities.

6. **Reproduction:** In cellular reproduction, the parent cell divides to form identical two daughter cells that may be used for growth in multicellular organisms.

7. **Excretion:** The process of removing wastes, which may be carried out by the lungs, the digestive system, the urinary system, or the integumentary system.

8. **Differentiation**: Specialization of embryonic tissues into different body tissues with limited capabilities. This occurs as different set of genes are switched off in the chromosomes of different cells.

1.1.6 Homeostasis

Homeostasis is very important to an organism. When homeostasis fails, it leads to illness or even death. Two general mechanisms are involved in homeostatic regulation: autoregulation and extrinsic regulation.

1. **Autoregulation (intrinsic regulation)**: Cells, tissues, and organ system have the ability to regulate their functions automatically in response to changes in the environment; e.g., when the O_2 level declines in a tissue, the cells release chemicals to dilate local blood vessels.

2. **Extrinsic regulation** results from the activities of the nervous system or endocrine system; for instance, after exercise, the nervous system increases your heart rate in order for blood to circulate faster.

A homeostatic regulatory mechanism consists of three parts:

1. **Receptor** is a sensor that is sensitive to a particular environmental change or stimulus.
2. **Control center** receives and processes the information supplied by the receptor.
3. **Effector** is a cell or organ that responds to the commands of the control center. Its activity either opposes or enhances the stimulus.

1.1.7 The Role of Negative Feedback

Most activities in our body are regulated by negative feedback mechanisms. An important example is the control of body temperature, a process called thermoregulation. The following steps are involved:

- A **stimulus** produces a response that opposes or negates the original stimulus—e.g., the temperature changes.
- A **Receptor** senses the stimulus (temperature sensors in skin and hypothalamus).
- **Control center** (in this case the thermoregulatory center in the brain) is affected, then sends commands to
- **Effectors**: Sweat glands in the skin increase secretions, and blood vessels in the skin dilate.

- **Response**: With increased heat loss, body temperature drops and normal temperature is restored.

1.1.8 The Role of Positive Feedback

A positive feedback mechanism occurs when the stimulus produces a reaction that exaggerates or enhances the effect of the stimulus rather than reverses it. Steps involved in the positive feedback in blood clotting are as follows:
1. A break in a blood vessel wall causes bleeding.
2. Damaged cells release chemicals.
3. Clotting begins.
4. Additional chemicals are released (causing clotting to accelerate).
5. A blood clot plugs the break in the vessel wall, and bleeding stops.

"Positive feedback is important in accelerating processes that must proceed to completion rapidly. Think of a positive feedback as a domino effect—once the domino begins to fall, the chain reaction causes each successive domino to fall until completion." (*Instructor's Guide Resource Guide* by Lucia Tranel and Alice Mills—*Fundamentals of Anatomy & Physiology* by Frederic H Martini, Third Edition)

1.1.9 Anatomical Position

A person is in an anatomical position when the following conditions are fulfilled:
1. Erect
2. Facing observer
3. Eyes forward
4. Arms at sides
5. Palms forward
6. Feet flat on the floor
7. Big toes together
8. Thumbs point away from the body
9. "Right" and "left" refer to the sides of the person being viewed

A person lying down in the anatomical position is said to be **supine** when face up, and **prone** when face down.

1.1.10 Common Anatomical Terms for Body Regions

Head (cephalic), consists of two regions:

1. Skull (cranial)

Common Term	Anatomical Term
Top of head	Parietal
Side	Temporal
Back	Occipital
Forehead	Frontal

2. Face (facial)

Frontal - forehead

Common Term	Anatomical Term
Eye	Ocular / *Orbital*
Nose	Nasal
Ear	Otic
Cheek	Buccal
Chin	Mental
Neck	Cervical

Patellar - kneecap

Pectoral - chest

Trunk: *Crural - leg*

Femoral - thigh
Antecubital - point of elbow
Coxal - hip

1. Front (ventral or anterior):

Common Term	Anatomical Term
Chest	Thoracic
Collar bone	Clavicular
Shoulder	Acromial
Breast bone	Sternal
Belly	Abdominal
Navel	Umbilical
Pelvis	Pelvic
Pubis	Pubic

Olecranon - point of elbow
Dorsum - back of hand

Nuchal - back of neck
Occipital - base of skull

2. Back (dorsal or posterior):

Common Term	Anatomical Term
Backbone - *Spinal column*	Vertebral
Loin	Lumbar
Tailbone	Sacral, Coccygeal

Gluteal — *buttock*
Perineal — *Perineum*

1.1.11 Planes

The principal planes include: midsagittal (median), Parasagittal, frontal (coronal), transverse (cross-sectional or horizontal), and oblique.

Sections are the surfaces resulting from cuts through body structures. They are named according to the plane on which the cut is made and include transverse, frontal, and midsagittal sections.

1.1.12 Directional Terms

Popliteal - hallow behind knee

Sural - calf
Plantar - sole
Calcaneal - heel

Common Terms	Anatomical Terms
Toward the upper part	Superior
Toward the lower part	Inferior, caudal
Front	Ventral or anterior
Back	Posterior or dorsal
Face up	Supine
Face down	Prone
Toward midline	Medial
Away from midline	Lateral
Near point of attachment	Proximal
Far from point of attachment	Distal
On the surface	Superficial
Away from body surface; more internal	Deep
On the wall of body cavity	Parietal
Directly on the organ	Visceral
Between	Intermediate

1.1.13 Body Cavities

The two principal body cavities are the dorsal and ventral cavities.
1. The dorsal cavity is subdivided into the cranial cavity, which contains the brain, and the vertebral (spinal) cavity, which contains the spinal cord.

2. The ventral cavity is subdivided by the diaphragm into an upper thoracic cavity and a lower abdominopelvic cavity

The thoracic cavity contains the pleural cavities and the mediastinum, which includes the pericardial cavity.
- The pleural cavities enclose the lungs, while the pericardial cavity surrounds the heart; these cavities are lined by serous membranes.
- The mediastinum is a mass of tissues between the lungs and contains the contents of the thoracic cavity except the lungs.

1.1.14 Abdominopelvic cavity
- Contains the peritoneal cavity, which is lined by the peritoneum
- The parietal peritoneum lines the inner surface of the body wall.
- The visceral peritoneum covers the enclosed organ.
- The narrow space between them contains small amounts of fluid.

Note: A few organs are not enclosed by the peritoneum. Such organs are referred to as retroperitoneal organs. Examples of such organs are the kidneys, the urinary bladder, and the pancreas. (These organs are "behind" the peritoneum.)

Note: The mediastinum is not a cavity, but rather a region within the thoracic cavity. The mediastinum contains the pericardial cavity.

1.1.15 The Nine Abdominopelvic Regions

Anatomists use two transverse and two parasagittal planes to divide the abdominopelvic cavity into nine regions.

1. The umbilical region is the centermost region deep to and surrounding the umbilicus.

2. The epigastric region is the region located superior to the umbilical region (above the belly).

3. The hypogastric (pubic) region is the region located inferior to the umbilical region (below the belly).

4. The right and left iliac, or inguinal, regions are the regions located lateral to the hypogastric region.

5. The right and left lumbar regions are the regions that lie lateral to the umbilical regions.

6. The right and left hypochondriac regions are the regions that flank the epigastric region laterally.

Right hypochondriac region	Epigastric region	Left hypochondriac region
Right lumbar region	Umbilical region	Left lumbar region
Right iliac (inguinal region)	Hypogastric (pubic) region	Left iliac (inguinal) region

1.2 Summary of the Body's Organ Systems

1.2.1 Integumentary

- Component: Hair, nails, and skin
- Functions: Perception, protection, and vitamin D production

1.2.2 Skeletal System

- Components: Bones, joints, and cartilages
- Functions: Protects and supports body organs
  Provides a framework the muscles use to cause movement
  Stores minerals
  Houses hemopoiesis

1.2.3 Muscular System

- Components: Skeletal system. Consists of muscles attached to skeleton by tendons
- Functions: Allows manipulation of the environment
  Locomotion
  Posture maintenance
  Production of heat

1.2.4 Lymphatic System

- Components: Spleen, thymus, lymphatic vessels, tonsils, and lymph nodes
- Functions: Defends against infection and disease (immunity)
                  Returns tissue fluids to the blood
                  Disposes of debris in the lymphatic stream
                  Houses white blood cells (lymphocytes) involved in immunity

1.2.5 Respiratory System
- Components: Sinuses, nasal conchae, larynx, trachea, bronchi, lungs, and alveoli
- Functions: Delivers oxygen and carbon dioxide to alveoli for exchange
                  Keeps blood constantly supplied with oxygen and removes carbon dioxide
                  Produces sounds for communication

1.2.6 Digestive System
- Components: Alimentary canal, mouth, oral cavity, esophagus, stomach, small
                  intestine, large intestine, rectum, and anus
- Accessory organs: Tongue, teeth, liver, salivary gland, pancreas, and gall bladder
- Functions: Break food down into absorbable units that enters the blood for distribution to
                  cells

1.2.7 Nervous System
- Components: Spinal cord, brain, special sense organs, and nerves
- Functions: Fast-acting control system of the body
                  Regulates activities of the body by detecting and responding to changes and
                  activating appropriate muscles and glands

1.2.8 Urinary System
- Components: Kidneys, ureters, urinary bladder, and urethra
- Functions: Eliminates nitrogenous wastes from the body
                  Regulates water, electrolyte, and acid–base balance of the blood

1.2.9 Endocrine System
- Components: Pituitary gland, thymus gland, pancreas, adrenals, gonads, and other
                  endocrine tissues in other systems
- Functions: Directs long-term changes in the activities of other organ systems
                  Adjusts metabolic activity and energy use by the body
                  Controls many structural and functional changes during development

1.2.10 Cardiovascular System
- Components: Heart, blood, and blood vessels
- Functions: Blood vessels transport blood, which carries oxygen, carbon dioxide,
                  nutrients, and wastes
                  Heat distribution and temperature regulation
                  The heart pumps blood

1.2.11 Male Reproductive System
- Components: Testes, epididymis, ductus deferens, ejaculatory duct, urethra, penis, and
                  glands
- Functions: Sperm and male sex hormone production
                  Delivery of sperm to
                  the female

1.2.12 Female Reproductive System
- Components: Ovaries, uterus, fallopian tube, vagina, and accessory glands

- Functions: Ovaries produce eggs and female sex hormones. Site of fertilization and development of the fetus
  Mammary glands produce milk to nourish the newborn.

Review Questions

1. The branch of biological science that deals with the function of organs and organ system is called _physiology_.
2. The tendency for physiological systems to stabilize internal conditions is called _homeostasis_.
3. Homeostatic regulation usually involves a(n) _receptor_ that is sensitive to a particular stimulus and a(n) _effector_ whose activity has an effect on the same stimulus.
4. _Autoregulation_ occurs when activities of cells, tissues, organs, or systems change intrinsically.
5. _Extrinsic Regula_ results from the activities of the nervous or endocrine systems.
6. In _positive_ feedback, the initial stimulus produces a response that escalates or exaggerates the stimulus.
7. When homeostatic mechanisms fail, an individual will experience the symptom of _disease_.
8. A person lying face down in the anatomical position is in the _prone_ position.
9. A cut parallel to the midsagittal plane would produce a(n) _parasaggital_ section.
10. The common name for the olecranon is the _elbow_.
11. Visceral pericardium is located _on heart_.
12. The mediastinum separates the _pericardial sac_ and contains the _pericardial_ cavity.
13. The diaphragm divides the ventral body cavity into a superior _thoracic_ and an inferior cavity. _abdominopelvic_
14. The serous membrane covering the stomach and most of the intestines is called the _peritoneum_.
15. The kidneys and urinary bladder are organs of the _urinary_ system.
16. The wrist is _distal_ to the elbow.
17. The chest is _superior_ to the umblicus.
18. The study of early developmental processes is termed _embryology_.
19. Skin, hair, and nails are associated with the _integumentary_ system.
20. Defense against infection and disease is the function of the _lymphatic_ system.

Chapter 2
The Chemical Level of Organization

Chemistry is the science that deals with the structure of matter.

2.1.1 Matter and States of Matter

Matters are substances, such as elements and compounds.

1. They have mass and occupy space.
2. They can neither be created nor destroyed but can be converted from one form to another.
3. Matter exists in solid, liquid, and gaseous states.

Energy is the capacity to do work.

2.1.2 Types of Energy

1. Potential is stored energy, inactive energy that has the potential to do work (e.g., batteries, dammed water, your leg muscles).
2. Kinetic is energy in action; it does work by moving objects.

2.1.3 Forms of Energy

1. Chemical energy is stored in the bonds of chemical substances (e.g., chemical energy in food is stored in the bonds of ATP),
2. Electrical energy reflects the movement of charged particles—e.g., nerve impulses.
3. Mechanical energy involves the movement of matter.
4. Radiant or electromagnetic is energy that travels in waves (visible light, infrared waves, radio waves, etc.).

2.2.1 Atoms

Atoms are the smallest stable unit of matter. They are composed of subatomic particles: protons, neutrons, and electrons.

1. Protons (p+): bear a positive electrical charge.
2. Neutrons (n or n°): are electrically neutral, or uncharged.
3. Electrons (e⁻): are much lighter than protons—1/1836 as massive—with a negative electrical charge.

2.2.2 Atomic Structure

Atoms normally contain equal numbers of protons and electrons.

2.2.3 Nucleus

1. Protons are located in the nucleus.
2. All atoms other than hydrogen have both neutrons and protons in their nuclei.
3. The number of protons in an atom is known as the atomic number.

2.2.4 Electron cloud

1. Electrons orbit the nucleus at high speed, forming an electron cloud.
2. It is much less dense than the nucleus.
3. It contains electrons located in discrete energy levels Atoms are most stable if the outermost energy level is filled. Chemical reactions occur to make atoms more stable.
4. Atomic measurements are reported in nanometers (10^{-9}).

2.3 Elements and Isotopes
2.3.1 Elements

1. Each element includes all atoms that have the same number of protons (atomic number).
2. Ninety-two elements exist in nature.
3. Elements cannot be changed or broken down into simpler substances in chemical reaction.
4. The human body contains 13 elements and 13 trace elements.

2.3.2 Isotopes

Isotopes are atoms whose nuclei contain the same number of protons, but different numbers of neutrons (they have identical chemical properties).

2.3.4 Mass number

Mass number is the total number of protons plus neutrons in the nucleus (used to designate isotopes). Examples include ^1H (hydrogen-1), ^2H (hydrogen-2, also called deuterium), ^3H (hydrogen-3, also called tritium).

2.3.5 Radioisotopes

Radioisotopes are isotopes whose nuclei spontaneously emit subatomic particles or radiation.

2.3.6 Atomic Weights

1. The atomic weight is the actual mass of an atom.
2. It is expressed in daltons or atomic mass units (amu).
3. It is the average mass number that reflects the proportion of different isotopes.

2.4 Molecules and Mixtures

1. Molecules are combination of two or more atoms held together by chemical bonds—e.g., hydrogen molecule (H_2), oxygen molecule (O_2), sulfur molecule (S_8).
2. A compound is a combination of two or more different kinds of atoms—e.g., water (H_2O), methane (CH_4).

2.4.1 Mixtures are substances that are composed of two or more components that are physically mixed. There are three types of mixtures:

1. **Solutions**: Homogeneous mixtures of components that may be solid, liquid, or gases—e.g., rubbing alcohol, which contains water (solvent) and alcohol (solute). Solvents are usually liquids present in a greater amount than solute.
2. **Colloids (emulsions)** are heterogeneous mixtures that often appear translucent. Solutes in colloids are larger than those in solutions but do not settle. The solutes scatter light (e.g., cytosol in cells).
3. **Suspensions** are heterogeneous mixtures with large, often visible solutes that tend to settle out (e.g., sand and water, blood).

2.5 Electrons and Energy Levels

1. Atoms are electrically neutral; every positively charged proton is balanced by a negatively charged electron.
2. Electrons occupy an orderly series of energy levels within the electron cloud.
3. The first electron shell corresponds to the lowest energy level.
4. The first energy level can hold a maximum of two electrons, the next eight electrons.
5. The first energy level is filled before any electrons enter the second; the second level is filled before the third, and so on.

6. The number of electrons in the outermost energy level determines the chemical properties of the element.
7. Atoms with unfilled energy levels will react with other atoms in a way to fill the outer shell.
8. Atoms with filled outermost energy levels are stable and do not readily react with other atoms (inert atoms such as helium, neon, and argon).

2.5.1 Chemical Bonds

Reactive elements, i.e. elements with unfilled outermost energy levels, readily interact or combine with other atoms to achieve stability by gaining, losing, or sharing electrons to fill their outermost energy level. There are three basic types of chemical bonds: ionic bonds, covalent bonds, and hydrogen bonds. Chemical bonds result in the creations of molecules and compounds.

Ionic bonds involve the transfer of up to three electrons.
1. Loss of electrons: cations are formed (+ charge).
2. Gain of electrons: anions are formed (- charge).
3. Ionic compounds contain anions and cations.

Examples of ionic compounds include salts and electrolytes (water-soluble compounds).

2.5.2 Covalent bonds involve the sharing of electrons between atoms.
1. Sharing one electron forms a single bond (no charge).
2. Sharing two electrons forms a double bond.
3. Sharing three electrons forms a triple bond.
4. Unequal sharing forms polar covalent bonds (partial charge).
5. Examples of covalent compounds are fats and oils (water insoluble).
6. Examples of polar covalent bonds are acids and sugars (water soluble).
7. Hydrogen bonds are weak covalent bonds between hydrogen (H) and nitrogen (N) or oxygen (O).

2.5.3 Molecular Weight
1. The molecular weight of a molecule is the sum of the atomic weights of its component atoms.
2. The molecular weight of a molecule in grams is equal to the weight of one mole of the molecule.

2.6 Chemical Reactions

1. Chemical bonds are made or broken to fill the outer energy levels with electrons.
2. There are collisions between compounds during chemical reactions.
3. The frequency of collisions is affected by speed, concentration, orientation of molecules, and activation energy.
4. A chemical reaction involves three participants:
   - **Reactants** are starting materials.
   - **Catalysts** speed up reactions but are not consumed.
   - **Products** are ending materials.

Types of Chemical Reactions

2.6.1 Decomposition reactions involve the breaking of molecules into smaller fragments.
1. $AB \rightarrow A + B$
2. Decomposition reactions involving water are called hydrolysis.
   $A\text{-}B\text{-}C\text{-}D\text{-}E\text{-}F + H_2O \rightarrow A\text{-} B\text{-}C\text{-}D\text{-}H + HO\text{-}E\text{-}F.$

2.6.2 A **synthesis reaction** is the opposite of decomposition (i.e., it puts together). It is the assembling of smaller molecules into larger molecules.
1. $A + B \rightarrow AB$
2. Formation of new bonds
3. Two substances becoming one, such as $Na + Cl \rightarrow NaCl$

2.6.3 Dehydration synthesis (**condensation)** is the formation of a complex molecule by the removal of water
1. $A\text{-}B\text{-}C\text{-}H + HO\text{-}D\text{-}E \rightarrow A\text{-}B\text{-}C\text{-}D\text{-}E + H_2O$
2. It is the opposite of hydrolysis reaction (anabolism).

2.6.4 Exchange reactions happen when parts of the reacting molecules are shuffled around to produce new products.
1. $AB + CD \rightarrow AD + CB$. Atoms change partners.
2. Bonds break; new bonds are formed.
3. $NaCl + KI \rightarrow NaI + KCl$

2.6.5 In **reversible reactions**, two reactions are occurring simultaneously, one a synthesis $(A + B \rightarrow AB)$ and the other decomposition $(AB \rightarrow A + B)$.
1. $A + B \leftrightarrow AB$
2. They are always indicated by a double arrow.
3. They convert products to reactants.

2.7 Inorganic Compounds

2.7.1 Water is the most abundant and important inorganic compound in the body (60 to 80 percent). Water is a universal solvent. Water has the following properties:
1. High heat capacity (absorbs and releases large amounts of heat before changing appreciably in temperature itself)
2. Has polar solvent properties (i.e., is a universal solvent)
3. Is a participant in chemical reactions such as hydrolysis, dehydration, and synthesis
4. Forms a resilient cushion around certain organs
5. Has a pH of 7.0

2.7.2 Salts are ionic compounds containing cations and anions other than hydroxyl ions (OH⁻). They are the products of acid–base reactions. They dissociate into component ions in water and are an essential source of electrolytes

2.7.3 Acids can react with many metals. They dissociate in water into hydrogen ions and anions.
1. Acids are proton donors with pH less than 7.0.
2. $HCl \rightarrow H^+ + Cl^-$

2.7.4 Bases dissociate in water, liberating hydroxyl ions and cations.
1. Bases are proton acceptors with pH greater than 7.0.
2. $NaOH \rightarrow Na^+ + OH^-$

2.7.4 pH: Acid–Base Concentration
1. The more hydrogen ions in a solution, the more acidic the solution is.
2. The pH scale runs from 0 to 14 and is logarithmic; a unit change = 10X change in concentration.
3. pH less than 7.0 is acid, pH greater than 7.0 is basic, pH of 7.0 is neutral.
4. pH of blood is 7.35–7.45.

2.7.5 Homeostatic Control of pH
1. Chemoreceptors detect pH changes in the blood.
2. pH is maintained within narrow limits by use of buffers.
3. Buffers are made from weak acids and the salts of weak acid (act as weak bases).
4. Buffers convert strong acids and bases to weak ones.
5. Physiological buffers include bicarbonate (extracellular) and phosphate (intracellular).

2.8 Organic Compounds
2.8.1 Carbohydrates
Carbohydrates are compounds that contain C, H, and O in a 1:2:1 ratio.
Classes
1. Monosaccharides: glucose, galactose, fructose, ribose, and deoxyribose
2. Disaccharides: sucrose, maltose, and lactose
3. Polysaccharides: starch, cellulose, and glycogen

Uses
1. For energy, glucose provides immediate energy, glycogen stored energy.
2. Carbohydrates are part of self-identification when combined with proteins and lipids.
3. Carbohydrates can be converted to other substances.

2.8.2 Lipids
Lipids contain mostly C and H with some O and sometimes P.
Classes
Triglycerides:
1. Composed of fatty acids and glycerol
2. Can be saturated (only single bonds between carbon atoms)
   - Monounsaturated: one double bond between carbon atoms
   - Polyunsaturated: more than one double bond between carbons

Steroids are flat molecules made of four interlocking hydrocarbon rings.

1. They are fat soluble and contain little oxygen.
2. Cholesterol found in cell membranes is an important steroid.
3. Cholesterol is also a raw material for vitamin D synthesis, steroid hormones, and bile salts.

Eicosanoids are derived from a 20-carbon fatty acid (arachidonic acid) found in cell membranes. Examples are prostaglandins and leukotrienes, which play important roles in blood clotting, inflammation, and labor contractions.

Note: Lipids may combine with proteins to form lipoproteins (HDL and LDL).

2.8.3 Proteins

1. Proteins contain C, H, O, N, and sometimes S.
2. They make up to 10 to 30 percent of cell mass.
3. They are the basic structural material of the body.

Components

1. Amino acids with carboxyl group (COOH) and amino group (NH$_2$)
2. A central carbon atom
3. A variable group known as an R group or side chain
4. Amino acids joined together by peptide bonds

Classes

1. **Structural:** Hair and nails
2. **Regulatory:** Hormones and neurotransmitters
3. **Contractile:** Muscles
4. **Immunological:** Antigens, interleukins, and antibodies
5. **Transport:** Hemoglobin and lipoproteins
6. **Catalytic:** Enzymes

Levels of Organization

1. **Primary:** Sequence of amino acids along the length of a single polypeptide
2. **Secondary:** Results from bonds between atoms at different parts of the polypeptide chain
3. **Folding pattern:** Alpha-helix and pleated sheet
4. **Tertiary:** Complex coiling and folding that gives protein a final three-dimensional shape
5. **Quaternary:** Interaction between individual polypeptide chains to form a protein complex

Fibrous and Globular Proteins

Fibrous

1. They are extended and strand-like.
2. Most exhibit only secondary structures.
3. They are insoluble in water and very stable, providing tensile strength and mechanical support.
4. They are the building materials of the body.

Globular

1. They are compact and spherical.
2. They exhibit tertiary and quaternary structures.
3. They are water soluble and play important roles in most biological processes.

Protein denaturation

Hydrogen bonds begin to break when the pH drops or the temperature rises above normal (physiological) levels. Proteins unfold and lose their specific three-dimensional shape.

Enzymes

1. Enzymes are large protein molecules.
2. They are biological catalysts.
3. They lower activation energy, thus allowing reactions to occur at normal body temperature.
4. Enzymes bind with substrates at active sites (induced-fit model).
5. Enzymes are specific, efficient, and tightly regulated.

Note: Some functional enzymes referred to as **holoenzymes** consist of two parts: an **apoenzyme** (which is the protein portion) and a **cofactor**. A cofactor can be made either of ions of metals such as **copper or iron** or of an organic molecule. Most organic molecules are derived from proteins, such as vitamin B complex.

2.8.4 Nucleic acids

Nucleic acids are large organic molecules that contain carbon, hydrogen, oxygen, nitrogen, nitrogen, and phosphorus.

1. **Deoxyribonucleic acid (DNA)** forms the genetic code inside each cell and thereby regulates most of the activities that take place in our cells.
2. **Ribonucleic acid (RNA)** relays instructions from the genes in the cell's nucleus to guide each cell's assembly of amino acids and proteins by the ribosomes.

Adenosine Triphosphate (ATP)

1. **Adenosine triphosphate (ATP)** is the principal energy-storing molecule in the body
2. It provides energy for muscular contractions, chromosome movement during cell division, cytoplasmic movement within cells, membrane transport processes, and synthesis reactions.

Review Questions

1. The center of an atom is called <u>its</u>_____.
2. Electrons in an atom occupy an orderly series of electron shells or _____.
3. A(n) _____ is a combination of two or more atoms and has different physical and chemical properties than individual atoms.
4. Ions with a positive charge are called _____.
5. Ions with a negative charge are called _____.
6. Chemical reaction that release energy are called _____.
7. Chemical reactions that require energy are called _____.
8. _____ control the rate of chemical reactions that occur in the human body.
9. _____ molecules are compounds that contain carbon as the primary structural atom.
10. A (n) _____ is a homogeneous mixture containing a solvent and a solute.
11. _____ are soluble inorganic compounds whose ions will conduct an electric current in solutions.
12. Molecules that readily dissolve in water are called _____.
13. If an isotope of oxygen has 8 protons, 10 neutrons, and 8 electrons, its mass number is _____.

14. Carbohydrates, lipids, and proteins are classified as _____ molecules.
15. When atoms complete their outer shell by sharing electrons, they form _____ bonds.
16. The bond between sodium and chloride in the compound sodium chloride (Nacl) is _____.
17. The most important high-energy compound in cells is _____.
18. A nucleotide consists of _____, _____, and _____.
19. Nucleic acids are composed of units called _____.
20. Each amino acid differs from another in the _____.
21. You would expect a peptide bond to link two _____.
22. Most of the fat found in the human body is in the form of _____.
23. The group of organic compounds containing carbon, hydrogen, and oxygen in a near 1:2:1 ratio is defined as a _____.
24. Molecules that have the same molecular formula but different structural formulas are called _____.
25. AB → A + B is to decomposition as A + B → AB is to _____.
26. The chemical behavior of an atom is determined by _____.
27. Isotopes of an element differ in the number of _____.
28. Atomic number represents the number of _____.
29. Radioisotopes have unstable _____.
30. The simplest chemical units of matter are _____.

Chapter 3
Cells: The Living Units

3.1 The Cell

The cell is the basic structural and functional unit of living organisms.

3.1.1 Plasma Membrane

Structure

1. It is made up of a phospholipid bilayer with a polar hydrophilic head (exterior) and a nonpolar hydrophobic tail (interior).
2. It contains cholesterol between the phospholipids that determine the fluid nature of the membrane.
3. It contains carbohydrates in the form of glycolipids and glycoproteins on the outside (glycocalyx).
4. Integral proteins penetrate deeply into the lipid bilayer, acting as channel proteins.
5. Peripheral proteins are attached to inner or outer surfaces.

3.1.2 The Fluid Mosaic Model

The fluid mosaic model of membrane structure depicts the plasma membrane as an extremely thin (7–8 nm) structure composed of a double layer, or bilayer, of lipid molecules with protein molecules dispersed in it. (Human Anatomy and Physiology by Elaine N Marieb, 5th Edition, page 68)

** visualize it as "hundreds of ping-pong balls floating on the surface of a pool of water with the occasional larger whiffle ball or foam rubber ball interspersed within the continuous sheet of ping- pong balls." (*Instructors' Resource Guide* by Lucia Tranel and Alice Mills, *Fundamentals of Anatomy and Physiology* by Fredric H Martini, Third Edition, page 22)

3.1.3 Roles of Membrane Proteins

1. **Receptors proteins or glycoproteins** with an exposed receptor site on the outer surface, which can attach to specific chemical signals (e.g., ligands such as calcium ions or hormones)
2. **Transport proteins** may provide a hydrophilic channel across the membrane that is selective for a particular solute. Some hydrolyze ATP as an energy source for active transport.
3. **Receptors linked to channel proteins** help form ligand-gated ion channels. Channel can open or close in response to chemical signals or ligands.
4. **Receptors linked to G protein complexes** alter the activity of a G protein complex located on the inner surface of the plasma membrane, causing ion channels to open.
5. **Cell-to-cell recognition** serves as identification tags that are specifically recognized by other cells.
6. **Channel proteins** are nongated ion channels that are always open, allowing permeability in cells at rest. Gated ion channels can be opened or closed by stimuli. Some, ligand ion channels, respond to ligands (proteins and glycoproteins), and some respond to changes in charge across the membrane (voltage-gated ion channels)

3.1.4 Carrier Proteins

1. **Uniporters** move one particle.
2. **Symporters** move two particles in the same direction simultaneously.

3 **Antiporters** move two particles in opposite direction at the same time.

3.2 Movement Across Cell Membranes

The following factors affect permeability of materials across the cell membrane:

1. Molecular size
2. Lipid solubility
3. Ionic charge
4. Carrier molecules

3.3 Types of Movement across the cell membrane

3.3.1 Passive Movement involves movement of materials across the plasma membrane without using energy. The process depends on the concentration gradient of substances and their kinetic energy.

3.3.1a Simple Diffusion is the movement of solutes from an area of higher concentration to lower concentration.

1. Concentration gradient is required. Kinetic energy is the energy source.
2. Solutes move from an area of higher concentration to lower concentration.

The following substances pass through lipid bilayer by simple diffusion: fat-soluble vitamins (A, D, E, and K), small nitrogenous waste molecules, and gases.

The following substances can diffuse through pores that can open or close in response to chemical or electrical signals: K, Na, Cl, bicarbonate ions, and water.

3.3.1b Osmosis is the movement of water through selectively permeable membranes from an area of low concentration of solute to an area of high concentration of solute.

Two factors are involved in osmosis.

- Osmotic pressure is the force with which pure water moves into that solution as a result of solute concentration or the force required to prevent water from moving across a membrane
- Osmolarity is the total solute concentration in an aqueous solution.

Tonicity

This is the term used to describe the effects of various osmotic solutions on cells.

1. Hypertonic: crenation (cell shrinks)
2. Isotonic: no osmotic flow of water into or out of the cell
3. Hypotonic: hemolysis (lysis)

3.3.1c Filtration

1. Utilizes hydrostatic pressure
2. Occurs in the kidneys, where blood is filtered

3.3.1d Facilitated diffusion

1. This is necessary if the substance is very large or very lipid insoluble.
2. It requires transporter proteins.
3. It is used to move large, water-soluble molecules or electrically charged molecules across membrane.

Three factors are involved:
1. concentration gradient
2. number of transporters
3. rate of combination of particle

3.4 Active Processes
1. needed to move materials against concentration gradients
2. required to move ions, amino acids, and monosaccharides

3.4.1 ActiveTransport
There are two types of active transports.

- ### Primary transport
1. It uses energy from ATP directly.
2. The sodium-potassium exchange pump exchanges intracellular sodium for extra-cellular potassium (for each ATP molecule consumed, three sodium ions are ejected and two potassium ions are reclaimed).

- ### Secondary transport
1. It uses energy stored in ion gradients (the concentration gradient for one substance provides the driving force needed by the carrier protein, and the second substance gets a "free ride").
2. The speed depends on Na gradient.

3.5 Bulk Transport (Vesicular Transport)
There are two types of bulk transport.

3.5.1 Endocytosis
1. **Phagocytosis** is cell eating—e.g., white blood cell engulfing bacteria.
2. **Pinocytocysis** is cell drinking (formation of endosomes filled with extracellular fluid).
3. **Receptor mediated endocytosis** is the main mechanism for the specific uptake of most macromolecules by body cells. Substances taken up by this method include enzymes, insulin, and LDL. Flu viruses and diphtheria toxin use this route to enter cells.

3.5.2 Exocytosis
1. Exocytosis is the functional reverse of endocytosis.
2. Secretary vesicles are formed, such as in the release of neurotransmitters and hormones, secretion of milk by mammary glands, and secretion of mucous by salivary glands.

3.6 Cytoplasm
1. Cytoplasm is the cellular material between the plasma membrane and the nucleus.
2. It is the site of most cellular activities.
3. It is composed of cytosol, organelles, and inclusions (glycogen granules, lipid droplets, and pigment granules).

3.7 Cytoskeleton is responsible for the movement of whole cells or organelles. Examples:
1. **Microtubules** are hollow tubes built from the globular protein tubulin, which gives strength to the cell.
2. **Intermediate filaments** strengthen the cell and help maintain shape. They also stabilize positions of organelle.
3. **Microfilaments** are composed of the protein actin. They anchor the cytoskeleton to integral

proteins of the cell membrane. They produce movement by interacting with myosin.

3.8 Organelles

Organelles are the internal structures that perform most of the tasks that keep the cell alive and functioning normally. They play specific roles in cell growth, maintenance, repair, and control.

1. **Microvilli** are finger-shaped projections of the cell membrane on their exposed surfaces.
2. **Centrioles** are cylindrical structures composed of short microtubules. They form spindle fibers associated with the movement of DNA strands during cell division.
3. **Cilia** are slender extensions of the cell membrane used for movement of materials past the cell.
4. **Flagella** are few and long; they are used for movement of cell.
5. **Ribosomes** function in protein synthesis. There are two types:
    1. Fixed for extracellular protein manufacturing (attached to endoplasmic reticulum)
    2. Free for intracellular protein manufacturing (scattered throughout the cytoplasm)
6. **Proteasomes** remove proteins produced by free ribosomes in the cytoplasm. They also remove and recycle damaged or denatured proteins.
7. **Endoplasmic reticulum (ER)** is a network of intracellular membranes connected to the nuclear envelope.
    There are two types:
        a. Agranular (smooth) ER: Lipid synthesis. Calcium storage and release and detoxification
        b. Granular (rough) ER: protein synthesis and modification.
8. **The Golgi apparatus** looks like a stack of dinner plates. It consists of five or six flattened membranous discs called cisternae. It processes, sorts, and packages lipids and proteins.
9. **Lysosomes** are digestive enzymes. Tay-Sachs disease is as result of lack of lysosomal enzymes.
10. **Peroxisome** protects the cell from the damaging effects of free radicals produced during catabolism.
11. **Mitochondria** are double- membraned; the outer surrounds the organelle, and the inner membrane contains numerous folds called cristae. They function in cellular respiration. They require oxygen to generate ATP.

The Nucleus

The nucleus is the largest and most conspicuous structure in a cell. It serves as a control center for cellular operations.
Contents of the nucleus:

1. The contents of the nucleus are called nucleoplasm.
2. The nucleoplasm contains ions, enzymes, and nucleotides required for making DNA and RNA

Nucleoli

1. Dark-staining spherical bodies found within the nucleus
2. Act as site for ribosome production

3.9 Gene Action
3.9.1 DNA

1. DNA is confined to the nucleus. It contains all genetic information required to run the cell.
2. It is copied exactly in normal cell division by the process of replication.
3. The DNA molecule winds around the histone, forming a complex known as a nucleosome. The nucleosomes are loosely coiled within the nucleus, forming the chromatin. Just before cell division, the coiling become tighter, forming chromosomes (23 pairs in humans)

3.9.2 The Transcription of RNA
1. Transcription is the process by which genetic information encoded in DNA is copied onto a strand of RNA called messenger RNA (mRNA).
2. DNA also synthesizes ribosomal RNA (rRNA) and transfer RNA (tRNA).

3.9.3 Translation
Translation is the process whereby information in the nitrogenous base sequence of the mRNA specifies the amino acid sequence of a protein.
1. It occurs on ribosomes in cytoplasm.
2. mRNA threads through grooves in ribosomes.
3. mRNA reads in three base sequences called codons.
4. Transfer RNA brings amino acids to ribosomes.
5. If the tRNA anticodon complements the mRNA codon, amino acid is deposited.
6. Peptide bonds form, and mRNA shifts.

Note: All nucleated cells except germ cells have the full complement of DNA.
Differentiation occurs during development when some segments of DNA are turned off in some cells while those segments remain "on" in other cells.
The rate of protein synthesis varies depending upon chemical signals that reach the cell.

3.10 Normal Cell Division
Cell division is the process by which cells reproduce themselves. Cell division involves two processes: nuclear division (mitosis and meiosis) and cytoplasmic division (cytokinesis).
1. **Somatic cell division** is the cell division that results in an increase in body cells, which involves nuclear division (mitosis) and cytokinesis.
2. **Reproductive cell division** results in the production of sperm and eggs (gametes).

3.10.1 Somatic Cell Division
1. During interphase, DNA molecules replicate themselves so that the exact copy of the chromosomal complement can be passed on to the daughter cells.
2. **The Cell cycle** is the series of changes a cell goes through from the time it is formed until it reproduces itself. The cell cycle encompasses two major periods: interphase, in which the cell grows and carries on its usual activities, and cell division.

Interphase (metabolic phase) is the period between cell division when a cell is carrying on every life process except division. Interphase is divided into G1, S, and G2 subphases.

1. **In the S** (for synthesis) phase, chromosomes are replicated, ensuring that the two future cells will receive identical copies of the genetic materials.
2. **The G1** (for gap or growth) phase is the first portion of interphase when the cell is engaged in growth, metabolism, and synthesis of proteins required for cell division.
3. **The G2-phase** is the period when enzymes and other proteins needed for cell division are

synthesized and moved to their proper sites.

4. **Replication** is the process whereby the DNA uncoils, allowing each exposed base pair to pick up a new complementary base.

Mitosis (nuclear division) is the series of events that parcel out the replicated DNA of the mother cell to two daughter cells. The process is divided into four stages: prophase, metaphase, anaphase, and telophase.

1. During **prophase**, the chromatin condenses and shortens into chromosomes, and the nucleus disassembles.
2. During **metaphase**, the chromatids align at the equatorial plane or metaphase plate.
3. At **anaphase**, the chromatids split at the centromere, and the two sister chromatids begin to move toward opposite poles of the cell.
4. At **telophase**, chromosome movement stops; the identical sets of chromosomes at the opposite poles of the cell uncoil and revert to their threadlike protein form, microtubules disappear, and a new nuclear envelope forms.

Cytokinesis (cytoplasmic division) begins during the late telophase and is completed after mitosis ends.

Review Questions

1 The main divisions of the cytoplasm are _____ and _____.

2. Masses of insoluble material that are sometimes found in cytosol are known as _____.

3. _____ cells are all of the cells of the body except the reproductive cells (sperm and ovum.

4. Receptor molecules on the surface of cells bind specific molecules called _____.

5. The transmembrane potential in an undisturbed cell is called _____.

6. Cellular reproduction of somatic cells is known as _____.

7. _____ is the process of duplicating DNA prior to cell division.

8. The process by which cells become specialized is called _____.

9. Endoplasmic reticulum is responsible for _____.

10. A unit in messenger RNA containing of a set of three consecutive nucleotides is termed a(n) _____.

11. White blood cells engulfing disease-causing bacteria illustrates _____.

12. Two types of vesicular transport include _____ and _____.

13. Cell shrinkage is to _____ as cell bursting is to _____.

14. The genetically controlled death of cells is called _____.

15. Crenation occurs when a blood cell is placed in a(n) _____ solution.

16. Synthesis of lipids takes place at the _____._____.

17. A process that requires cellular energy to move a substance against its concentration gradient is called _____.

18. A solution that contains a lower solute concentration than the cytoplasm of a cell is called _____.

19. Define the term diffusion:

_____

20. The principal cations in the body fluids are _____ and _____.

Chapter 4
The Tissue Level of Organization

Tissues are collections of cells (with a common embryonic origin) and cell products that perform a specific but limited range of functions. A branch of anatomy that deals with the study of tissues is called histology.

4.1 Epithelial Tissues

Epithelial tissues perform the functions of covering, lining, and secreting (glands).
Characteristics:
1. The cells are closely packed.
2. The cells may be in a single layer or multiple layers.
3. Polarity: apical (exposed) and basal (attached)
4. Many cell junctions are present.
5. There are no blood vessels (avascular).
6. The cells are attached to underlying tissues by basement membranes.
7. Innervated (supplied by nerve fibers)
8. Highly mitotic (high capacity to regenerate)
9. Epithelial tissue arises from the three primary germ layers.

Functions: Protection, filtration, secretion, digestion, absorption, excretion, sensation

4.1.1 Cell Junctions

Cell junctions are points of contact between adjacent plasma membranes. There are three types of cell junctions:
1. **Tight junctions** forms fluid-tight seals between cells and are common among epithelial cells of the stomach, intestines, and urinary bladder. They prevent leakage.
2. **Anchoring junctions** fasten cells to one another or to extracellular material.
   - They are common in tissues subjected to friction and stretching (e.g., outer layer of skin, muscle of heart, neck of uterus, lining of gastrointestinal tract).
   - Examples include **desmosomes**, **hemidesmosomes**, and **adherens** junctions.
3. **Communicating junctions** allow the rapid spread of action potentials from one cell to the next in the heart muscle, parts of the nervous system, and the gastrointestinal tract.

   Note: Many cancer cells lack gap junctions between them, leading to the uncoordinated growth of these cells.

4.2 Classification of Epithelial tissues

4.2.1 Simple squamous epithelium consists of a single layer of flat, scale-like cells.
- It is adapted for diffusion and filtration and is found in lungs and kidneys; in serous membranes, it functions in osmosis and secretion.
- It is referred to as endothelium in the lining of the heart and blood vessels.
- Mesothelium lines the thoracic and abdominopelvic cavities.

4.2.2 Simple cuboidal epithelium consists of a single layer of cube-shaped cells.
- It is adapted for secretion and release of substances such as mucus, perspiration, or enzymes. It also plays a role in the absorption of substances.
- It is found in the ovaries, anterior surface of lens capsules, kidney tubules, the choroid plexus

of the brain, and the lining of terminal bronchioles of the lungs.

4.2.3 Simple columnar, nonciliated epithelium
- It consists of a single layer of nonciliated, rectangular cells.
- It functions in the secretion of mucus and in the absorption in the kidneys.
- It lines the gastrointestinal tract, where it performs the function of absorption through fingerlike projections called microvilli. The goblet cells perform the function of secretion.

4.2.4 Simple columnar, ciliated epithelium
- Its rectangular-shaped cells are ciliated.
- It moves fluids or particles along a passageway by ciliary action.
- It is found in the upper respiratory tract, fallopian tubes, uterus, sinuses, and the central canal of the spinal cord.

4.2.5 Stratified squamous epithelium
- It consists of several layers of cells in which the most superficial cells are flattened.
- It performs the function of protection.
  Types
    1. Keratinized squamous epithelium forms the outer layer of the skin. It contains water-proof keratin that is resistant to friction.
    2. The nonkeratinized variety is found in the mouth, esophagus, vagina, part of the epiglottis, and the tongue.

4.2.6 Stratified cuboidal epithelium
- It consists of several layers of cells in which the top layer is cube-shaped.
- It functions in protection.
- It is located in the ducts of the male urethra and ducts of adult sweat gland.

4.2.7 Stratified columnar epithelium
- It consists of several layers of cells in which the top layer is rectangular.
- It functions in protection and secretion.
- It can be found in portions of the male urethra and in the excretory ducts of some glands.

4.2.8 Transitional epithelium
- It consists of several layers of cells with variable appearance depending on the degree of stretching.
- It is capable of stretching, allowing for distension of an organ.
- It lines the urinary bladder and portions of ureter and the urethra.

4.2.9 Pseudostratified columnar epithelium
- It consists of one layer but gives the appearance of several layers. All cells are resting on the basement membrane.
- It functions in secretion and movement of mucus by ciliary action.
- It is found in the excretory ducts of large glands, part of the male urethra, and the Eustachian tubes. The ciliated variety contains goblet cells and is mostly found in the upper respiratory tract.

4.3 Glandular Epithelium

These are glands with specialized cells or groups of cells that secrete substances into ducts, onto surfaces, or into blood. Glands are epithelium with a supporting network of connective tissue.
Types:
1. Exocrine glands secrete their products into ducts that empty at the surface of covering and lining epithelium or directly onto free surfaces.
   - Examples are sweat glands, sebaceous glands, and digestive glands.
   - The functional classification are as follows:
     a) The holocrine gland accumulates its secretion in the cytosol and is discharged along with the dead cell. The dead cell is replaced by a new one.
     b) Merocrine glands discharge their secretion by the process of exocytosis.

Examples are the sweat glands, the pancreas, and the salivary gland.
- c) Apocrine glands accumulate their products at the apical surface. A portion pinches off from the rest of the cell to form the secretion. The remaining part of the cell repairs itself.

2. Endocrine glands are ductless glands, and the product of their secretions is hormones, which enter extracellular fluid and diffuse into blood.

4.4 Connective Tissue
Characteristics:
1. It is a matrix: ground substance + fibers (determines tissue fluidity)
2. It has widely separated cells.
3. It is not on exposed surfaces.
4. It has abundant blood supply.

4.4.1 Types of cells
1. Immature cells: Undifferentiated, their names end with –*blast*; they produce the matrix. **They are messenchyme cells.**
2. Differentiated, names end with –*cyte*; they maintain the matrix.

Examples:
1. **Macrophages** phagocytize or provide protection.
   - Fixed macrophages stay in position in connective tissue.
   - Wandering macrophages move by amoeboid movement through connective tissue.
2. **Plasma cells** produce antibodies.
3. **Mast cells** are found along small blood vessels, where they release histamine and heparin in response to injury.
4. **White blood cells (leukocytes)** respond to injury or infections.
5. **Adipocytes** are found in the dermis of the skin.

4.4.2 Ground substances
These are amorphous materials that fill the space between cells.

Examples:
1. **Chondroitin sulfate**
2. **Hyaluronic acid:** Polysaccharide that functions as lubricant
3. **Proteoglycans:** Protein and polysaccaride. The protein part attaches to hyaluronic acid, acting to trap large amount of water
4. **Adhesive molecules** hold proteoglycan aggregates together.

Functions:
1. Support and bind cells.
2. Act as a medium of exchange of materials.
3. Play a role cell differentiation, migration, division, and metabolism.
4. Enclose organs as a capsule and separate organs into layers.
5. Store fat.

4.4.3 Fibers

Fibers in the matrix provide strength and support for tissues. There are three types of fibers.

1. **Collagen fibers**, composed of the protein collagen, are very tough and resistant to stretching but allow some flexibility in tissue. They are found in bone, cartilage, tendons, and ligaments.
2. **Elastic fibers** consist of the protein called elastin, which provides strength and stretching in the skin, blood vessels, and the lungs.
3. **Reticular fibers** are made up of collagen and glycoprotein, which provide strong support around cells, nerve fibers, blood vessels, and skeletal and smooth muscle fibers. They are also components of basement membranes and the framework of many soft organs.

4.5 Loose Connective Tissue—with Special Properties

Type	Type of fibers	Type of cells	Location	Functions
1. Areolar	Collagen Elasticity Reticular	Macrophages Plasma cells Adipocytes Mast cells	Nerves Blood vessels Subcutaneous layer of skin Mucus membranes	Provides strength and elasticity
2. Adipose		Adipocytes	Renal fascia Subcutaneous skin layer Pericardium Orbital socket Yellow bone marrow	Support Protection Energy reserve Heat production and retention in newborns
3. Reticular	Reticular fibers	Reticular cells	Around muscle Reticular lamina Around blood vessels Organ stroma Hemopoietic tissues	It binds smooth muscle together. It forms organ stroma.

4.6 Dense Connective Tissue

Tissue Type	Traits (characteristics)	Locations	Functions
1. **Regular**	Shiny white matrix Mostly collagen fibes with fibroblasts between them	Tendons/aponeurosis	Provides strong attachments

2. **Irregular**	Collagen fibers are randomly arranged with few fibroblasts	Reticular dermis perichondrium periosteum Joint capsules Heart valves Fasciae	Provides strength
3. **Elastic**	Contains branching elastic fibers with fibroblasts in between them	Lung Respiratory tree True vocal cords Elastic arteries Ligaments between Vertebrae Suspensory ligament of penis	Permits distension of organs

4.7 Cartilage—with Special Properties

Type of cartilage	Traits (characteristics)	Locations	Functions
1. **Hyaline**	Most abundant type Contains shiny, bluish-white ground substance Contains many chondrocytes	Anterior ends of ribs Embryonic cartilage Ends of long bones Respiratory cartilages	Support Flexibility Allows movements at joints
2. **Fibrocartilage**	Contains scattered chondrocytes Contains bundles of collagen fibers	Pubic symphysis Menisci of knee Intervertebral discs	Support Fusion of bones

4.8 Bone Tissue

Types:
1. Spongy: contains **trabeculae** and red bone marrow
2. Compact: made up of osteons (Haversian system), which is the structural unit of compact bone
   Contains
      1. **Lamellae:** group of hollow tubes of bone matrix
      2. **Lacunae:** cavities at the junctions of lamellae
      3. **Osteocytes:** spider-shaped cells that occupy lamellae
      4. **Canaliculi:** hairlike canals that connect the lacunae to each other and to the central canal

Note: Connective tissue that surrounds bone is called **periosteum**.

Functions:
1. Support
2. Protection
3. Leverage

4. Housing hemopoietic tissue
5. Storage of fat and minerals

4.9 Blood

Plasma: Liquid part (matrix) contains water, salts, and gases.

Cells:

1. **Erythrocytes** (red blood cells)
2. **Leukocytes** (white blood cells)
3. **Thrombocytes** (platelets) (fragments)

Functions:
1. Blood clotting
2. Immunity
3. Transport of nutrients, gases, and wastes

4.10 Epithelial Membranes

The combination of an epithelial layer and an underlying connective tissue constitutes an epithelial membrane. Examples are mucous, serous, and cutaneous membranes.

1. **Mucous membranes (mucosae)** line cavities that open to the exterior, such as the reproductive tract and gastrointestinal tract.
    a. The epithelial layer of a mucous membrane plays an important role in the body's defense mechanisms by acting as a barrier to pathogens
    b. The connective tissue layer of a mucous membrane is called lamina propria.
2. A **serous membrane (serosa)** lines a body cavity that does not open directly to the exterior and covers organs such as the lung (pleura), the heart (pericardium), and the digestive organs and reproductive organs (peritoneum).
    a. The **parietal peritoneum** lines the cavity.
    b. The **visceral peritoneum** makes contact with organ.
    c. The epithelial layer secretes a lubricating serous fluid that reduces friction between organs and the walls of the cavities in which they are located.
3. The **cutaneous membrane** is the skin.
4. **Synovial membranes** line joint cavities, bursae, and tendon sheaths. They are not typical epithelial membranes since they lack epithelium. They secrete synovial fluid in joints.

4.11 Tissue Repair
4.11.1 Basic Description
- Tissue repair requires that cells divide and migrate.
- It is initiated by growth factors released by injured cells.
- It restores homeostasis.
- It requires cell division of differentiated and undifferentiated cells.
- It replaces dead or damaged cells.

4.11.2 Repair Processes
    a. If the injury is superficial, tissue repair involves pus removal, scab formation, and parenchymal regeneration.
    b. Major injury involves parenchyma and stroma. The fibroblasts divide rapidly, and

granulation tissue is formed. Scarring remains after the repair.

4.11.3 Factors Affecting Repair
1. **Nutritional status:** Proteins; vitamins A, B, C, D, E, and K; and carbohydrates
2. **Circulation**
3. **Age**
4. **Race**

Note: Regeneration is the replacement of destroyed tissue with the same kind of tissue.
Fibrosis involves the proliferation of fibrous connective tissue called scar tissue.

Related Clinical Terms
1. **Adenoma:** Any neoplasm of glandular epithelium, benign or malignant. The malignant type is more specifically called adenocarcinoma.
2. **Carcinoma:** Cancer arising in an epithelium; accounts for 90 percent of human cancers.
3. **Keloid:** Abnormal proliferation of connective tissue during healing of skin wounds; results in large, unsightly mass of scar tissue at the skin surface.
4. **Lesion:** Any injury, wound, or infection that affects tissue over an area of a definite size
5. **Marfan's syndrome:** Genetic disease resulting in abnormalities of connective tissues due to defects in fibrillin
6. **Sarcoma:** Cancer arising in the mesenchyme-derived tissue, that is, in connective tissue and muscle

Embryonic Tissue
The ectoderm, mesoderm, and endoderm are called the germ layers because they give rise to all tissues of the body.
1. **Endoderm:** the inner layer that forms the digestive tract and its derivatives
2. **Mesoderm:** the middle layer that forms tissues like muscle, bone, and blood vessels
3. **Ectoderm:** the outer layer that forms the skin
4. **Neuroectoderm:** a portion of the ectoderm that becomes the nervous system
5. **Neural crest cells:** group of cells that break away from the neuroectoderm. They give rise to parts of the peripheral nerves, skin pigment, the medulla of the adrenal gland, and many tissues of the face.

Review Questions
1. The epithelium that line the body cavities is known as _____.
2. The lining of the heart and blood vessels is called_____.
3. _____ is(are) the fluid component of connective tissue.
4. The combination of fibers and ground substance in supporting connective tissues is known as _____.
5. The watery ground substance of blood is called _____.
6. Epithelia and connective tissues combine to form _____ that cover and protect other structures and tissues in the body.
7. Interstitial fluid that enters a lymphatic vessel is termed _____.
8. Intercalated discs and pacemaker cells are characteristic of _____.
9. The four types of tissues in the body are _____. _____,

_____, and _____.

10. _____ attach skeletal muscles to bones, and _____ connect one bone to another.

11. The three types of connective tissues include _____, _____, and _____.

12. Antibodies are produced by _____.

13. The pancreas is an example of a(n) _____ gland.

14. Watery perspiration is an example of a(n) _____ secretion

15. The muscle tissue that shows no striations is _____ muscle.

16. The serous membrane lining the abdominal cavity is the _____.

17. The membranes that line cavities that communicate with the exterior of the body are called _____.

18. The most common type of cartilage is _____ cartilage.

19. Chondrocytes are to cartilage as osteocytes are to _____.

20. Tissues that provide strength and support for areas subjected to stresses from many directions are _____.

21. Name the dominant fiber type in dense connective tissue: _____

22. Cells that store fat are called _____.

23. Stratified cuboidal epithelia would be found _____.

24. Simple cuboidal epithelia would be found _____.

25. Epithelium is connected to underlying connective tissue by _____.

Chapter 5
The Integumentary System

The integumentary system has two components: the **epidermis** (above or superficial) and the **dermis**, an underlying area of connective tissues. The organs that make up the integumentary system are the skin and its derivatives, such as hair, nails, glands, and nerve endings.

5.1.1 Epidermis

Layers
1. Stratum corneum is the exposed surface of the skin (several layers of keratinized cells).
2. Stratum lucidum is present in the thick skin of the palm and soles.
3. Stratum granulosum contains keratin and keratohyalin.
4. Stratum spinosum contains several of keratinocytes bound together by desmosomes.
5. Stratum basale (germinativum) is the innermost epidermal layer that forms the epidermal ridge and contains large basal cells that are highly mitotic.

5.1.2 Cells
1. Keratinocytes produce keratin, which gives protective properties to the skin.
2. Melanocytes synthesize melanin, which shields the skin from the damaging effect of UV. There is an equal number of melanocytes in every race.
3. Langerhans cells arise from bone marrow, macrophages that help to activate the immune system.
4. Merkel cells are associated with sensory nerve endings to form the Merkel disc, which functions as a sensory receptor for touch and superficial pressure.

Note
1. **Desquamate** is the sloughing off of the older cells as they are replaced by the deeper layers when they undergo mitosis

2. During **keratinization** the cells are filled with keratin as they move outward, die, and serve as a layer to resist abrasion.

5.1.3 Associated Structures
1. **Hair follicles**: Extend from the epidermal surface into the dermis (the deep end is expanded to form the hair bulb)
2. **Hair plexus**: Knot of sensory nerve endings around the hair bulb
3. **Hair papilla**: Portion of dermal tissue that contains capillaries that supply nutrients to the growing hair and signals it to grow
4. **Arrector pili** muscle is a bundle of smooth muscle associated with each hair follicle; it pulls the hair into an upright position. It forms "goose bumps."
5. **Sebaceous glands** are alveolar glands that secrete sebum. They are present all over body except on the palms and soles.
6. **Nails** are scale-like modifications of the epidermis.

5.2 Dermis
The dermis has two major components: a superficial papillary layer and a deeper reticular layer.

5.2.1 Outer Papillary Layer

1. Contains **areolar connective tissue, elastic fibers, capillaries, Meissner's corpuscles, free nerve endings** and **papilla**
2. Dermal **papillae**: peg-like projections on the superficial part of papillary layer
3. Epidermal ridges are present in the palms and soles where they increase friction. They are genetically determined and unique for each of us (they are what make our fingerprints).

5.2.2 Deeper Reticular Layer
1. Contains dense **irregular connective tissue, collagen,** and **elastic fibers**
2. End **organs of Ruffini,** a mechanoreceptor that responds to mechanical pressure
3. Blood vessel and sweat glands
4. It has variable thickness

5.3 Subcutaneous Layer
The connective tissue fibers of the reticular layer are extensively interwoven with those of the subcutaneous layer, or hypodermis.

Components:
1. Adipose tissue
2. **Pacinian corpuscles** (a mechanoreceptor) and nerves
3. Large arteries and veins at the superficial region

It is the site for subcutaneous injection with hypodermic needles.

5.4 Dermal Strength and Elasticity
Collagen fibers: provide strength but resist stretching; are easily bent or twisted
Elastic fibers: permit stretching and then recoil to the original length
Skin water content: helps skin flexibility and resilience (skin turgor)
Stretch marks: extensive distortion of the dermis that occurs over the abdomen during pregnancy

Lines of Cleavage
Most of the collagen and elastic fibers at any location are arranged in parallel bundles oriented to resist the forces applied to the skin during normal movement.

5.5 The Basis of Skin Pigmentation
5.5.1 Melanin
1. A polymer made of tyrosine that is present mostly in epidermis
2. It is more abundant in the extremities, genital areas, nipples, and face.
3. Clusters of melanin form **freckles** and **age spots**.
4. Approximately equal numbers in all races
5. Gives color range from yellow to black
6. The shade of color varies with amount of melanin.
7. Produced by **melanocytes**
8. Albinos and people afflicted with vitiligo do not have melanin.

5.5.2 Carotene
1. Located mostly in dermis, stratum corneum, and subcutaneous layer, confers yellow to orange color to the skin

2. Most obvious in the palms and soles

5.5.3 Hemoglobin

1. In red blood cells in capillaries
2. Gives pink to red hue color to skin

Clinical Notes

1. Overall skin color is combination of pigments.
2. Homeostatic imbalance causes a change in skin color, such as cyanosis, jaundice, erythema, or hematoma.
3. **Mongolian spots:** blue-gray patches
4. **Addison's disease:** Pituitary gland secretes large quantities of ACTH
5. Some tumors affecting the pituitary gland result in the secretion of large amount of MSH,causing a darkening of the skin.
6. **Basal cell carcinoma:** Cancer of the germinative cells in the stratum basale of the skin
7. **Melanoma:** A skin cancer originating in the melanocytes
8. **Keloid** is a thickened area of scar tissue covered by a shiny, smooth epidermal surface.
9. **Urticaria (hives)** is an extensive dermatitis resulting from an allergic reaction to food, drugs, and insect bites.
10. A **scab** is a blood clot that forms at the surface of wound to the skin.
11. **Decubitis ulcers (bedsores)** are ulcers that occur in areas subjected to restricted circulation.

5.6 The Roles of the Epidermal Growth Factor

It has widespread effects on epithelia throughout the body. It is produced by the salivary glands and glands of the duodenum. Following are some of its function:

1. promoting the divisions of germinative cells in the stratum germinativum
2. accelerating the production of keratin in keratinocytes
3. stimulating epidermal development and repair
4. stimulating synthetic activity and secretion by epithelial glands

5.6.1 Skin Physiology

The general functions of the skin and subcutaneous layer include the following:

1. **Regulation of body temperature**
   a. Sweat glands—evaporation of sweat changes the temperature
   b. peripheral blood flow
   c. Adipose tissue in the subcutaneous layer insulates
   d. Hair insulates
2. **Protection:** Provides physical barrier against abrasion, infectious agents, UVradiation,and dehydration.
3. **Sensation:** Nerve endings and receptors to detect temperature, touch, pressure, and pain
4. **Excretion:** Skin eliminates water, salts, organic compounds, and heat.
5. **Immunity:** Skin provides nonspecific immunity through the help of sebum and lysozyme.
6. **Blood reservoir:** Skin stores 8 to 10 percent of the body's total blood.

7. **Vitamin D synthesis**: Precursor molecule in skin activated by exposure to the sun

5.7 Hair and Hair Follicles

- Distribution: Scalp, axilla, and pubis
- Hair is an epidermal derivative
- Hair protects scalp from injury, sun, and heat loss
- Protects respiratory tract and ear from foreign particles
- Hair growth is affected by illness, diet, blood loss, hormonal state, and cancer therapy.

5.7.1 Anatomy of Hair

Hair projects above the surface of the skin almost everywhere except the soles of the feet and the palms of the hands.

1. **Shaft**: Projects from the skin
2. **Root:** The portion of hair embedded in the skin
3. **Medulla**: Central core of the hair that consists of large cells and hair spaces (absent in fine hairs)
4. **Cortex**: Surrounds the medulla and contains several layers of flattened cells
5. **Cuticle**: Outermost layer formed from a single layer of cells, most keratinized layer
6. **Hair follicle**: Extends from the epidermal surface into the dermis
7. **External root sheath**
8. **Internal root sheath**
9. **Bulb**: Expanded deep end of the hair follicle
10. **Hair papilla**: A nipple-like bit of the epidermal tissue that contains capillaries that supply nutrients to the hair
11. **Matrix**: Actively dividing area of the hair bulb

5.7.2 Sebaceous Glands (Oil Glands)

1. They are holocrine glands present all over the body except on the palms and soles.
2. Discharge oily liquids into hair follicles
3. Secreted product formed is called sebum (mixture of triglycerides, cholesterol, proteins, and salts).
4. Functional parts lie in the dermis.
5. Located predominantly in areas requiring lubrication
6. Functions include lubrication, water retention, and serving as an antibacterial agent.
7. Accumulations form blackheads, pimples, and acne.

5.7.3 Sudoriferous (Sweat) Glands

The skin contains two types of sweat glands: excrine glands and apocrine glands.

Excrine glands are more numerous.

1. Secretory part found in subcutaneous layer
2. Excrete sweat through pores and skin surface
3. Produces higher water content sweat
4. Function throughout life

5.7.4 Apocrine Glands

1. Located in the axilla, pubis and breast areolae
2. Secretory part located in the dermis or subcutaneous layer
3. Secrete into hair follicle
4. Activated during puberty

5. Stimulated by emotional stress and sexual excitement
6. Basis for body odor when secretions are decomposed by bacteria

5.7.5 Ceruminous Glands

1. These are modified sweat glands that are found in the ear.
2. They lie in the subcutaneous layer.
3. They excrete onto the surface of the external auditory canal and ducts of sebaceous glands.
4. They produce cerumen (ear wax), which provides a physical barrier.

5.8 Epidermal Wound Healing

1. Epidermal wounds include abrasions or first-degree or second-degree burns with the center of the wound extending into the dermis and the edges involving only superficial damage to the epidermal cells.

2. Epidermal wounds are repaired by the enlargement and migration of basal cells to fill the wound. The movement of the cells is stopped by **contact inhibition.**

5.8.1 Steps in Deep Wound Healing

An injury that extends deep into the epidermis is referred to as a deep wound. Its repair is more complex than epidermal healing and often results into scar formation.

1. In the **inflammatory phase**, blood vessels' permeability to white blood cells increases. The mesenchymal cells move to the site, forming fibroblasts. Blood clot unites the wound edges.
2. In the **migratory phase**, epithelial cells beneath the scab bridge the wound, fibroblasts begin to migrate along the fibers and synthesize scar tissue, and damaged blood vessels begin to grow. **Granulation tissue** begins to fill the wound.
3. During the **proliferative phase**, the epithelial cells grow under the scab, collagen fibers are randomly deposited, and blood vessel growth continues.
4. During the **maturation phase**, the scab comes off, the collagen fibers become more organized, and the blood vessels are restored to normal.

Scars (fibrosis)

There are three types of scars: (1) normal, (2) hypertrophied, and (3) keloid.

- Scars contain dense collagen fibers.
- No epidermis
- Contain fewer blood vessels

5.9 Thermoregulation

This involves homeostatic control to maintain body temperature.

5.9.1 Negative Feedback Loop

- Affectors: Thermoreceptors in skin
- Control center: Hypothalamus
- Effectors: (1) Sweat glands become more active if body is too warm, less active if body is too cool.
  (2) Blood vessels vasodilate if body is too warm, vasoconstrict if body is too cool.

5.10 Skin Aging

Aging affects all the components of the integumentary system. Effects of aging on the skin are as follows:

1. Most pronounced after late 40s
2. The epidermis thins as germinative cell activities decline, the dermis thins, and subQ fat

decreases.

3. Collagen fibers decrease in number, stiffen and break, and become tangled.
4. Fibroblasts numbers decrease.
5. Elastic fibers stiffen, clump, and fray.
6. Immune response decreases, leading to poor healing.
7. Sebaceous glands shrink, and skin and derivatives become brittle.
8. Growth of hair and nails slows.
9. Thermoregulation becomes less effective.
10. Melanocytes functioning decreases, leading to graying.
11. Melanocytes become bigger, causing "age spots."
12. Vasculature: thicker and less permeable capillaries

5.11 Burns

Burns are significant injuries in that they can damage the integrity of large areas of the skin.

5.11.1 Causes

1. Excessive heat
2. Excessive sun (UV) exposure
3. Corrosive chemicals
4. Electricity
5. Radioactivity

5.12 Types of Tissue Damage

1. Local: tissues are actually damaged.
2. Systemic: homeostasis is disrupted thereby causing dehydration, fluid shifts, oligouria, shock, circulatory changes, and infection.

Review Questions

1. List three accessory structures of the skin: _____. _____, and _____.

2. The two components of the cutaneous membrane are the _____ and _____.

3. The most abundant cells in the epidermis are _____.

4. The layer of the epidermis that contains cells undergoing division is the _____.

5. Name the epidermal layer found only in the skin of the palms of the hands and the soles of the feet: _____

6. The layer of the epidermis that contains melanocytes is the _____.

7. The layer of the skin that contains the blood vessels and nerves closest to the surface of the skin is the _____ layer.

8. The layer of the skin that contains bundles of collagen fibers and the protein elastin and is responsible for the strength of the skin is the _____ layer.

9. Glands that discharge an oily secretion into hair follicles are _____ glands.

10. Most body odor is the result of bacterial metabolism of the secretions produced by _____ glands.

11. Sensible perspiration is produced by _____ glands.

12. What is the function of the arrector pili muscle?

13. What causes stretch marks?

14. What causes scar tissue?

15. What causes wrinkles and sagging skin in elderly individuals? _____

16. The protein that permits stretching and recoiling of the skin is _____

17. What causes suntan?

18. Cyanosis is indicated by

19. Where is lanugo found?

20. What is the common cause of dandruff?

21. Deficiency or complete absence of skin pigment is called _____.

22. Patches of depigmented skin is called _____.

23. The cancer of epithelial tissue is called _____.

24. A cancerous tumor made up of pigmented cells is called _____.

25. Low body temperature is referred to as _____.

26. A second-degree burn is characterized by

27. The superficial outer layer of the epidermis is the _____.

28. The vascular and nutritive layer of the skin is the _____.

29. The pink color of the skin of Caucasians is due largely to

30. UV rays of the sun are absorbed by _____.

Chapter 6
Skeletal Tissue

Osseous tissue is a supporting connective tissue that contains specialized cells and a matrix consisting of extracellular protein fibers and ground substance.

6.1 The Bone Matrix

The matrix is made up of the following:

6.1.1 Mineral salts include the following:

The mineral salts in the matrix determine the bone consistency.

1. **Hydroxyapatite** (interaction of calcium phosphate and calcium hydroxide)
2. **Calcium carbonate**
3. Ions such as **sodium, magnesium**, and **fluoride**

6.1.2 Protein fibers include collagen fibers, which allow flexibility and tensile strength. Protein fibers make up to one-third of the weight of bone.

6.1.3 The Cells of Bones

Although osteocytes are most abundant, bone contains four types of cells: osteocytes, osteoblasts, osteoprogenitor cells, and osteoclasts.

1. **Osteoprogenitors** contain cells that produce daughter cells that differentiate into osteoblasts. They are located in the inner periosteum, endoderm, and blood vessels.
2. **Oteoblasts** function in bone building by secreting the bone matrix. They do not have mitotic potential.
3. **Osteocytes** are mature bone cells with no mitotic potential. These are actually osteoblasts that are surrounded by bone. They help to maintain the bone matrix.
4. **Osteoclasts** are derived from circulating monocytes and help in bone remodeling.

6.2 Cancellous and Compact Bones

Bones can be classified according to the amount of bone matrix relative to the amount of space present within the bone.

6.2.1 Cancellous Bone
- Consists of interconnecting rods or plates of bone called trabeculae
- The spaces between the trabeculae contain bone marrow and blood vessels.
- Trabeculae consist of several lamellae with osteocytes located in lacunae between lamellae.
- Trabeculae are oriented along the lines of stress within a bone.

6.2.2 Compact Bone
- Denser and has fewer spaces than cancellous bone
- Lamellae of compact bone are oriented around the blood vessels.
- Vessels that run parallel to the long axis of the bones are contained within central, or Haversian, canals lined with endosteum.
- Concentric lamellae are circular layers of bone matrix that surround a central canal.
- Osteon (Haversian system) is a single central canal, its contents, and associated lamellae.
- Perforating (Volkmann's) canals run perpendicular to the long axis of the bone.

6.3 Structure of a Long Bone
- The **diaphysis (shaft)** is composed of mostly compact bone.

- The end of a long bone is mostly cancellous bone with outer layer of compact bone.
- The end of a long bone is covered with hyaline cartilage (**articular cartilage**).
- The primary ossification center is in the diaphysis.
- The **epiphysis** is the part of the long bone that develops from a center of ossification.
- **Epiphyseal** or the **growth plate** separates the epiphysis from the diaphysis.
- The **epiphyseal line** is formed when the epiphyseal plate becomes ossified when the bone stops growing in length.
- The **medullary cavity** is a large internal space along the diaphysis of a long bone filled with marrow.

Note: Immature bones are fetal bones and adult repairing bones. They tend to have more cells that are randomly arranged. Matured bones cells are less random and contain more matrix.

6.4 Ossification (Osteogenesis)

The process of replacing other tissues with bone is called ossification. Two major forms of ossification exist: **endochondral** and **intramembranous**.

6.4.1 Endochondral (Intracartilaginous) Ossification

This refers to the formation of bone within the hyaline cartilage model.

Stages

The development of the cartilage model by the differentiation of mesenchymes into chondroblasts occurs in the following stages:

Growth of Cartilage
- Cartilage grows in length (interstitial growth).
- Cartilage grows in diameter (appositional growth).
- Cartilage cells in the center of the diaphysis accumulate glycogen, enlarge, and burst, releasing chemicals that alter pH and trigger the calcification of cartilage.
- Chondrocytes die because they are deprived of nutrients, forming space within the cartilage model.
- The nutrient artery enters the perichondrium of the cartilage model, stimulating the perichnodrial cells to form osteoblasts.
- A periosteal bone collar forms around the diaphysis.
- Periosteum develops from perichondrium.

Primary Ossification Center Forms
- Capillaries grow into spaces, and osteoblasts deposit bone matrix over disintegrating calcified cartilage. In this way spongy bone is forming within the diaphysis at the primary ossification center.
- Osteoclasts break down the newly formed spongy bone in the center of the bone, thereby leaving the medullary cavity.
- Compact bone and diaphyses form.

Epiphysis Develops
- Secondary ossification centers develop in epiphyses, forming spongy bone around the time of birth.
- Hyaline cartilage remains as the epiphyseal plate for as long as the bone growth continues.

6.4.2 Intramembranous Ossification

This begins when osteoblasts differentiate within mesenchymal or fibrous connective tissue. The steps in the process of intramembranous ossification can be summarized as follows:

- Step 1: Mesenchymal cells aggregate, differentiate, and begin the ossification process. The bone expands as a series of spicules spread into surrounding tissues.
- Step 2: As the spicules interconnect, they trap blood vessels within the bone.
- Step 3: Over time, the bone assumes the structure of spongy bone. Areas of spongy bone may later be removed, creating marrow cavities. Through remodeling, spongy bone formed in this way can be converted to compact bone.

Note: Flat bones of the skull, the mandible, and the clavicle are formed by this process.

6.4.3 How Osteoprogenitors Become Cartilage or Bone

- Oteoprogenitors become chondroblasts if capillaries are absent. The chondroblasts begin to make cartilage.
- Osteoprogenitors become osteoblasts if capillaries are present. The osteoblasts make bone via either endochondral growth or intramembranous growth.

6.5 Bone Remodeling

The organic and mineral components of the bone matrix are continuously being recycled and renewed through the process of remodeling.

- New bone replaces old bone.
- Move matrix to reinforce stress lines
- Repair worn and injured bone
- Mobilize mineral stores of calcium and phosphate
- Balance bone construction and destruction: High activity of osteoblasts leads to Paget's disease, while hyperactivity of osteoclasts leads to osteoporosis.

6.5.1 Minerals Required

- **Calcium** is part of the component matrix.
- **Phosphorus** is part of the components of the matrix.
- **Magnesium** is involved in osteoblast activity.
- **Boron** reduces Ca loss and increases estrogen levels.
- **Manganese** is required to put down bone.

6.5.2 Vitamins Required

- **D**: Calcitrol increases absorption of calcium. It also limits Ca loss from urine.
- **C**: Ascorbic acid maintains matrix. It is required for collagen growth and repair.
- **A**: controls developmental activity of osteoblast and osteoclast

6.5.3 Hormonal Effects

Normal bone growth and maintenance depend on a combination of nutritional and hormonal factors.
1. **Human Growth hormone (hGH)** for general growth
2. **Sex hormone** (**testosterone** and **estrogen**) stimulate osteoblasts to produce faster than the

rate at which epiphyseal cartilage expands. Over time, the epiphyseal plate narrows and eventually closes.

3. **Insulin** increases metabolism and glucose levels.
4. **T3/T4** involved in metabolism and protein synthesis
5. **Calcitonin:** Calcium balance
6. **Parathyroid hormone:** Calcium balance

6.6 Fractures

Despite its mineral strength, bone can crack or even break if subjected to extreme loads, sudden impact, or stress from unusual directions.

6.6.1 Classifications and Types of Fractures

- Degree of separation: Partial (greenstick) or complete
- Degree of exposure:

  Simple: Fractures are completely internal.

  Compound: Fractures project through the skin.

6.6.2 Degree of Alignment

1. **Nondisplaced:** Retain the normal alignment of the bones or fragment
2. **Commuted:** Affected area is shattered into multitude of bony fragments.
3. **Displaced:** Produce new and abnormal bone arrangements

   Shape or direction: **Depressed, oblique, longitudinal, transverse Spira**, and **stellate** fissure

Specific Terms

- **Colle's fracture:** A break in the distal portion of the radius, resulting from reaching out to cushion a fall
- **Pott's fracture:** Occurs at the ankle and affects both bones of the leg
- **Hangman's fracture:** A broken neck involving a fracture of an upper cervical vertebra similar to injury suffered by hanging; always involves the axis

6.7 Fracture Repair

Fractures heal even after severe damage, provided that the blood supply and the cellular components of the endosteum and periosteum survive. Repair may take months and involves gradual mineral deposition and decreased blood supply. Steps in the repair process are as follows:

6.7.1 Fracture Hematoma

1. **Hematoma formation.** Blood vessels and periosteum are damaged during fracture, leading to hemorrhage. As a result, a **hematoma (mass of blood clot)** forms at the fracture site. The bone cells around the area begin to die due to lack of nutrients, and the tissue at the site become swollen, painful, and inflamed.
2. **Fibrocartilaginous callus formation.** Within days, soft calus (granulation tissue) begin to form. New capillaries begin to grow into the hematoma, and phagocytic cells invade the area and begin to clean up the debris. Fibroblasts from the surrounding peritoneum migrate into the area and begin to synthesize collagen fibers. Some of them differentiate into chondroblast to lay down new matrix while others become osteoblasts that form spongy bone.
3. **Bony callus formation.** Osteoblasts begin to make new bone trabeculae in the fibrocartilaginous callus and gradually convert it to **bony (hard) callus** of spongy bone. This process begins three to four weeks after injury and may last up to two to three months.

4. **Bone remodeling**. The excess material on the bone shaft exterior and within the medullary cavity is removed, and compact bone is laid down to reconstruct the shaft walls.

6.8 Homeostasis and Calcium Balance

Calcium ion homeostasis is maintained by a pair of hormones with opposing effects. These hormones (parathyroid hormone and calcitonin) coordinate the storage and absorption of calcium ions.
1. Bone serves as a calcium reservoir and blood calcium buffer.
2. Calcium is required for muscle contraction, cardiac rhythm, nerve impulse conduction, formation of enzyme cofactors, and blood clotting.

Blood calcium levels are maintained at 8.5 – 10.3 mg/dL.

6.8.1 Hormonal Regulation of Calcium

6.8.1a Parathyroid hormone (PTH):
(1) It is produced in parathyroid glands
(2) acts on bone to raise Ca levels via negative feedback: cAMP increases rate of PTH release, PTH increases osteoclast activity, and calcium levels return to normal.
(3) acts on kidneys to raise blood Ca levels by decreasing Ca loss through urine. It increases phosphate loss through urine. With increased formation of calcitrol, calcium levels return to normal.

6.8.1b Calcitonin is produced in the thyroid gland. It lowers calcium levels by decreasing osteoclast activity and increasing loss of calcium through urine.

Review Questions

1   Mature bone cells are called _____.
2.   Cells that synthesize the organic components of the bone matrix are called

_____.
3.   Large, multinucleated cells that can dissolve the bony matrix are called

_____.
4.   In bone, calcium phosphate forms crystals of _____.
5.   The shaft of a long bone is called the _____.
6.   The head of a long bone is called the _____.
7.   The region of a long bone where the epiphysis meets the diaphysis is known as the

_____.
8.   _____ prevents damaging bone-to-bone contact within movable joints.
9.   During the process of _____, an existing tissue is replaced by bone.
10.   The process of depositing calcium into a tissue is called _____.
11.   Any projection or bump on a bone is termed a(n) _____.
12.   A smooth, grooved articular process shaped like a pulley is termed a(n)

_____.
13.   A(n) _____is a rounded passageway for passage of blood vessels and nerves.
14.   The adult skeleton contains _____ major bones.
15.   The process of producing new bone matrix is called _____.
16.   The matrix in spongy bone forms struts and plates called _____.
17.   A(n) _____ fracture is completely internal and does not break the skin.
18.   An open or _____ fracture projects through the skin.

19. In a(n) _____ fracture one side of the shaft is broken and the other side is bent.

20. A break in the distal portion of the radius is called a(n) _____ fracture.

21. Name the stem cell whose divisions produce osteoblasts.

    _____

22. Which part of the bone tissue stores adipose tissue for energy reserves?

    _____

23. Hypersecretion of parathyroid hormone would produce changes in the bone similar to those associated with _____.

24. Osteoclast activity is greater when calcitonin is present/absent (choose one)

25. Blood calcium levels is greater when parathyroid hormone is increased/decreased

26. Intramembranous ossification begins with the _____ cells.

27. The innermost part of the bone is known as the _____.

28. The functional units of mature compact bone are called _____.

29. Endochondral ossification begins with the formation of _____.

30. The lining of the marrow cavity is called the _____.

31. The central canal of an osteon contains _____.

Chapter 7

The Skeletal System: The Axial Skeleton

The axial skeleton forms the longitudinal axis of the body. The axial skeleton has 80 bones, roughly 40 percent of the bones in the body. The axial components are as follows:

1. The skull (eight cranial bones and 14 facial bones)
2. Bones associated with the skull (six auditory ossicles and the hyoid bone)
3. The vertebral column (24 vertebrae, the sacrum, and the coccyx)
4. The thoracic cartilage (the sternum and 24 ribs)

7.1 Cranial Bones

7.1.1 Frontal

The common features of the frontal
bones are:

- **Squama**: The most anterior part (forehead)
- **Supraorbital margin**: Thickened superior margins of the orbits that lie under the eyebrows
- **Supraorbital foramen (notch)** allows the supraorbital artery and nerve to pass to the forehead
- **Glabella**: The smooth portion of the frontal bone between the orbits
- **Frontal sinus**
- **Frontal suture:** The frontal suture articulates posteriorly with the paired parietal bones.
- **Orbit roof**

7.1.2 Parietal

- They form most of the superior and lateral aspect of the skull.
- The four largest sutures occur where the parietal bones articulate with other cranial bones.
    1. **The coronal suture**: Parietal bones meet the frontal bone anteriorly.
    2. **The sagittal suture**: The right and left parietal bones meet superiorly.
    3. **The lambdoid suture**: The parietal bones meet the occipital bone posteriorly.
    4. **The squamous (squamosal) suture**: Parietal and temporal bones meet on the lateral aspect of the skull.

7.1.3 Temporal

1. **Temporal squama**
2. **Zygomatic process** meets the zygomatic bone of the face anteriorly, forming the zygomatic.
3. **Petrous portion** contributes to the cranial base and unites with the sphenoid bone to form the middle cranial fossa, which supports the temporal lobe of the brain.
4. **Carotid foramen** transmits the internal carotid artery into the cranial cavity.
5. **Jugular foramen** allows passage of the internal jugular vein and three cranial nerves.
6. **Mandibular fossa** receives the condyle of the mandible, forming the temporomandibular joint.
7. **Articular tubercle**
8. **Mastoid process** is the anchoring site for neck muscles.
9. **Stylomastoid foramen** is located between the styloid and mastoid processes and allows

cranial nerve VII (facial nerve) to leave the skull.

10. **External acoustic meatus** allows sound to enter the ear through the canal.

11. **Styloid process**: An attachment point for several muscles of the tongue and neck and the ligament that secure the hyoid bone of the neck to the skull

7.1.4 Occipital

The occipital bone forms the walls of the posterior cranial fossa, which supports the cerebellum. Important features are:

1. **Foramen magnum**: "Large hole" through which the brain connects with the spinal cord.\
2. **Occipital condyles**: Articulate with the first vertebra and permits a nodding movement of the head
3. **Hypothalami canal (fossa)**: Allows the nerve of the same name to pass
4. **External occipital protuberance**: Knoblike projection
5. **Superior nuchal line**: Anchors many neck and back muscles, marks the upper limit of the inferior nuchal line
6. **Inferior nuchal line:** Also anchors the neck and back muscles

7.1.5 Sphenoid

The sphenoid bone is the keystone of the cranium. It forms the central wedge that articulates with all other cranial bones. Important features are:

1. The **body** contains the paired sphenoid sinuses.
2. The **sella turcica** is the saddle-shaped prominence on the superior surface of the body.
3. The **hypophyseal fossa** forms the enclosure for the pituitary gland (hypophysis).
4. **Greater wings** project laterally from the body.
5. **Lesser wings** form part of the floor of the anterior cranial fossa and part of the medial walls of the orbits.
6. The **optic foramen** allows the optic nerves to pass to the eyes.
7. The **superior orbital fissure** allows cranial nerves that control eye movements (III,VI) to enter the orbit.
8. **Pterygoid processes** anchor the pterygoid muscles important in chewing.

7.1.6 Ethmoid

This lies between the sphenoid bone and the nasal bones of the face. It is the most deeply situated bone of the skull. The important features are:

1. **Lateral masses** flank the perpendicular plate and contain ethmoid sinuses.
2. **Superior and middle nasal conchae (turbinates)** protrude into the nasal cavity.
3. A **perpendicular plate** divides the nasal cavity into left and right halves.
4. The **cribriform plate** is the superior surface of the ethmoid bone that forms the roof of the nasal cavities and the floor of the anterior cranial fossa.
5. The **crista galli** is the point of attachment for the outermost covering of the brain (dura mater).
6. **Olfactory foramina** are located on the cribriform plate. They allow the olfactory nerves to pass from the smell receptors in the nasal cavities to the brain.

7.2. Facial Bones

The facial skeleton is made up of 14 bones, of which only the mandible and the vomer are unpaired. The maxillae, zygomatics, nasals, lacrimals, palatines, and inferior conchae are paired bones.

7.2.1 Nasal
- They form the bridge of the nose They articulate with the frontal bone superiorly, the maxillary bones laterally, and the perpendicular plate of the ethmoid bone posteriorly.

7.2.2 Maxillae
The maxilla bone fuses medially. It forms the upper jaw and central portion of the facial skeleton. Important features are:

1. The **maxillary sinus** is the largest of the paranasal sinuses.
2. The **alveolar process** contains the upper teeth.
3. The **palatine process** forms the anterior two-thirds of the hard palate.
4. The **inferior orbital fissure** is located deep within the orbit. It permits the zygomatic nerves, the maxillary nerve, and blood vessels to pass to the face.
5. **Infraorbital foramen** allows the infraorbital nerve and artery to reach the face.

7.2.3 Zygomatic bones are commonly called the cheekbones.
- They articulate with the zygomatic processes of the temporal bones posteriorly, the zygomatic process of the frontal bone superiorly, and the zygomatic processes of the maxillae anteriorly.

7.2.4 Mandible
The mandible is the largest and strongest bone of the face. The important features are:
1. The **body** forms the chin.
2. **Ramus:** Each ramus meets the body posteriorly at a mandibular angle.
3. **Condylar process**
4. The **mandibular notch** separates the two processes arising from each ramus.
5. The **coronoid process** is the insertion point for the large temporalis muscle that elevates the lower jaw during chewing.
6. The **mandibular foramen** is located on the medial surface of each ramus and permits the nerves responsible for tooth sensation to pass to the teeth in the lower jaw.
7. The **alveolar process** contains the sockets in which the teeth are embedded.
8. The **mental foramen** is the opening on the lateral aspects of the mandibular body that allows blood vessels and nerves to pass to the skin of the chin and lower lip.

7.2.4 Lacrimal bones contribute to the medial walls of each orbit.
- Lacrimal fossa houses the lacrimal sac, part of the passageway that allows tears to drain from the eye surface into the nasal cavity.

7.2.5 Palatine is formed from two bony plates, horizontal and perpendicular. It contains the
- Inferior nasal concha
- Vomer

7.3 Paranasal Sinuses

- Function in mucus production, lighten the skull, and produce resonance

- Located on the frontal bone, maxillary bone, and sphenoidal bone

7.4 Fontanels

The largest fibrous areas between the cranial bones are known as fontanels.

Functions: Allow passage of head through birth canal and allow growth of brain during infancy

7.5 Vertebral Column

1. The adult vertebral column or spine consists of 26 bones: the vertebrae (24), the sacrum, and the coccyx or tail bone.
2. Between adjacent vertebrae, from the first cervical (atlas) to the sacrum, are intervertebral discs that form strong joints, that permit various movements of the vertebral column, and that absorb vertical shock.
3. The vertebral column contains four normal curves: two primary (thoracic and sacral) and two secondary (cervical and lumbar). These curves give strength, support, and balance.

7.6.1 Functions
- Acts as attachment site for ribs and back muscles

7.6.2 Vertebrae

Each vertebra consists of three basic parts: (1) the vertebral body, (2) the vertebral arch, and (3) articular processes (facets).

Note: The first cervical vertebral, the **atlas** (C1) has no body and no spinous process. The axis (C2) has a body, spine, and other typical processes. It has a knoblike **dens** or **odontoid process**.

7.7 The Appendicular Skeleton

The appendicular skeleton includes the bones of the limbs and the supporting elements, or girdles, that connect them to the trunk.

Types of Bone

a. **Long bones** are mostly compact bone (e.g., femur and humerus).
b. **Short bones** are mostly spongy bone (e.g., carpal and tarsal bones).
c. **Flat bones** are thin, spongy bones sandwiched between compact bones (e.g., cranium, ribs, sternum, and scapulae).
d. **Irregular bones** are bones with complex shapes, such as vertebrae and facial bones.
e. **Sesamoid bones** are small bones wrapped in tendons (e.g., patellae)
f. **Wormian bones** are sutural bones found in the joints of some cranial bones.

7.7.1 Pectoral (Shoulder) Girdle

a. The pectoral, or shoulder, girdle consists of the **clavicle** and the **scapula.** It attaches upper extremities to the trunk. There is no attachment to vertebrae, thereby allowing free movement of the upper extremities.
b. The clavicle is the most frequently broken bone in the body.
c. The scapulae articulate with other bones anteriorly.

7.8 Upper Extremity

Thirty separate bones form the skeletal framework of each upper limb.

7.8.1 Humerus (arm)

1. The **head** fits into the glenoid cavity of the scapula.
2. The **anatomical neck** marks the extent of the joint capsule.
3. **Greater tubercle**: lateral
4. **Lesser tubercle**: medial
5. The **intertubercular groove** guides a tendon of the biceps muscle of the arm to its attachment point at the rim of the glenoid cavity.
6. The **surgical neck** is the most frequently fractured part of the humerus.
7. **Body** (shaft)
8. The **deltoid tuberosity** is the attachment site for the deltoid muscle of the shoulder.
9. The **capitulum** articulates with the radius.
10. The **radial fossa** marks the course of the radial nerve
11. The **trochlea** articulates with the ulna.
12. The **coronoid fossa** is superior to the trochlea on the anterior surface.
13. The **olecranon fossa** is on the posterior surface and permits the ulna to move freely.
14. **Medial epicondyle** for muscle attachment

15. **Lateral epicondyle**

7.8.2 Radius and Ulna
(Forearm)

The forearm consists of two parallel long bones, the radius and the ulna, which form the skeleton of the forearm, or antebrachium.

7.8.2a The **ulna** is on the medial side.

Markings
1. Olecranon process
2. Coronoid process
3. Trochlear notch: grips the trochlea of the humerus, forming a hinge joint
4. Radial notch: point where the ulna articulates with the head of the radius
5. Head: distal end of the shaft
6. Styloid process: point where ligaments run to the wrist

7.8.2b The **radius** is located on the lateral side of the body.

Markings
1. Head
2. Radial tuberosity: anchors the biceps muscle of the arm
3. Neck
4. Styloid process: anchoring site for ligaments that run to the wrist
5. Ulnar notch: articulates with the ulna

7.8.3 Carpals (wrist) are made up of the following eight bones:
1. scaphoid
2. lunate
3. triquetrum
4. pisiform
5. trapezium

6. trapezoidcapitate
7. hamate

7.8.4 Metacarpals (palm)

7.8.5 Phalanges(fingers)

- Fingers are numbered 1 to 5 beginning with the thumb, or pollex
- Each finger has three phalanges: distal, middle, and proximal
- The thumb has no middle phalanx

7.9 Pelvic Girdle

The pelvic girdle attaches the lower limbs to the axial skeleton, transmits the weight of the upper body to the lower limbs, and supports the visceral organs of the pelvis. It is formed by a pair of hip bones each also called an os coxae or coxal bone.

- Posteriorly bound to sacrum
- Connected together anteriorly by pubis symphysis
- Contains three bones fused together

Description of the coxa bone

7.9.1 The **ilium** is the superior aspect of os coxa. It contains the following parts:
1. Iliac crest
2. Anterior superior iliac spine
3. Anterior inferior iliac spine
4. Posterior superior iliac spine
5. Posterior inferior iliac spine
6. Greater sciatic notch where nerves pass through to enter the spine
7. Sacroiliac joint

7.9.2 Spines are attachment points for muscles of the trunk, hip, and thigh.

7.9.3 The **ischium** forms the posteroinferior part of the hip bone. The major parts of the ischium are
1. Ischial spine
2. Lesser sciatic notch
3. Ischial tuberosity
4. Ramus
5. Obturator foramen

7.9.4 The **pubis or pubic bone** forms the anterior portion of the hip bone. The main parts are
1. Body
2. Pubic crest
3. Pubic tubercle, one of the attachments for the inguinal ligament
4. Obturator foramen, through which blood vessels and nerve pass
5. The bodies of the two pubic bones are joined by a fibrocartilage disc, forming the pubic symphysis joint.
6. The lateral aspect of contains deep fossa called acetabulum, which accepts the head of the femur.

7.10 Lower Extremity

The lower limbs carry the entire weight of the erect body. The three segments of each lower limb are the thigh, the leg, and the foot.

7.10.1 The **femur** or thighbone is the longest and heaviest bone in the body, just distal to the pelvic girdle.

7.10.2 The **patella (knee cap)**, a sesamoid bone, lies anterior to the knee joint. It increases the leverage of the tendon when the knee is flexed.

7.10.3 The **tibia (shinbone)** is medially located and bears the major portion of the weight of the leg. Important markings:

1. Lateral condyle
2. Medialcondyle
3. Intercondylar eminence
4. Tibial tuberosity
5. Medial malleolus
6. Fibular notch

7.10.4 The **fibula** is parallel to the tibia and laterally located. Major markings:

- Head
- Lateral malleolus, which forms the lateral ankle bulge and articulates with the talus

7.10.5 Tarsals constitute the ankle and share the weight associated with walking. The bones in this group:

- Talus
- Calcaneus (heel bone)
- Cuboid
- Navicular
- Anterior medial, intermediate, and lateral cuneiforms

7.10.6 Metatarsal bones consist of five small bones that are numbered 1 to 5 beginning on the medial (great toe) side of the foot.

Phalanges (toes) consist of 14 small bones. There are three phalanges in each digit except for the great toe, the **hallux**, which has only two.

Review Questions

Match the bone marking listed on the left with the appropriate bone listed on the right.

1.	Acromion	Tibia
2.	Greater trochanter	Coxal
3.	Leteral malleolus	Fibula
4.	Olecranon	Clavicle
5.	Ulnar notch	Radius
6.	Obturator foramen	Femur
7.	Conoid tubercle	Ulna
8.	Intercondylar eminence	Scapula

9. When one falls with outstretched arms, the radius may fracture transversely about one inch from the distal end. This particular injury is called a _____ fracture.

10. The opening defined by the inferior surfaces of the pubis, the inferior surfaces of the ischial tuberosities, and the coccyx is called _____.

11. A pollex is formed by two _____.

12. The knuckles of the hand are really the heads of the _____.

13. The head of the femur articulates with the _____.
14. The head of the humerus articulates with the _____.
15. The only sesamoid bone of the lower limb is the _____.
16. The term calcaneus refers to the _____.
17. The acromion is part of the _____.
18. A point of attachment of the two pubic bones is the _____.
19. The bony edge of the two pelvis bones is called the _____.
20. The space enclosed by the true pelvis is called _____.

Chapter 8

Joints and Articulations

Joints are classified by structure and function. Structurally there are fibrous, cartilaginous, and synovial joints. Functionally, there are synarthroses, amphiarthroses, and diarthroses.

8.1 Functional Classification of Articulation

8.1.1 Synarthroses (immovable)
(1) **Synostosis**: sutures in bony joints
(2) **Synchondrosis**: growth plate of hyaline cartilage (e.g., epiphyseal plate)
(3) **Gomphoses**: tooth in the socket

8.1.2 Amphiarthroses (slightly movable)
(1) **Syndesmoses** can be found at the distal tibia-fibula joint, forming fibrous connective tissue, interosseous membranes.
(2) **Symphysis** can be found between pubic bone and vertebrae.

8.1.3 Diarthroses (freely movable synovial joints)
(1) **Gliding (Arthroidal):** wrist, ankle
(2) **Hinge (Gynglymus):** elbow, knee, finger.. Hinge joints permit lexion and extension.
(3) **Pivot (Trochoid):** The joint between the atlas and the dens of the axis. It permits rotation of the head.
(4) **Condyloid (Elipsoidal):** The joint between radius and carpal bone; permits angular motions such as flexion and extension, abduction and adduction and circumduction.
(5) **Sellar (Saddle):** thumb: hand; circumduction
(6) **Sphroidal (Ball and socket):** shoulder, hip; permits universal movements (that is, in all axes and planes)

8.2 Common Types of Movements at Diarthroses
1. **Flexion** decreases the angle between articulating bones.
2. **Extension** increases the angle between articulating bones.
3. **Hyperflexion** is a continuation of extension beyond the anatomical position.
4. **Rotation**: A bone moves in a single plane around the longitudinal axis.
5. In **circumduction**, the distal end of a part of the body moves in a circle.
6. **Abduction** usually refers to movement of a bone away from the midline of the body.
7. **Adduction** usually refers to movement of a bone toward the midline of the body.

8.2.1 Special Movements at Diarthroses
1. **Elevation**: Upward movement
2. **Depression**: Downward movement
3. **Protraction**: Movement of the mandible or pectoral girdle forward on a parallel plane to the ground
4. **Retraction**: Movement of the mandible or pectoral girdle backward on a parallel plane to the ground
5. **Inversion**: Movement of the sole of the foot inward/medially

6. **Eversion**: Movement of the sole of the foot outward/laterally
7. **Dorsiflexion**: Bending of the foot in the direction of the dorsum
8. **Plantar flexion:** Movement of the foot in the direction of the plantar surface
9. **Supination**: Movement of the forearm that turns the palm of the hand anteriorly
10. **Pronation:** Movement of the forearm that turns the palm of the hand inferiorly

8.3 Joint Diseases

1. **Rheumatism** is a condition marked by pain in bones, ligaments, tendons, and muscle and includes arthritis, which is inflammation of joints.
2. **Rheumatoid arthritis** is an autoimmune disease that affects articular cartilage.
3. **Osteoarthritis** is a degenerative disease of the bone due to aging.
4. **Gouty arthritis** is the accumulation of uric acid in soft tissues.
5. **Lyme disease** is caused by bacterium transmitted by ticks.
6. **Bursitis** is the inflammation of bursae.
7. **Ankylosing spondylitis** is an inflammatory disease that affects the sacroiliac joint. It usually affects young men and may result in a loss of mobility.

Review Questions

1. An immovable joint is called _____.
2. A slightly movable joint is called _____.
3. A suture is an example of a(n) _____.
4. A joint formed by the fusion of two bones is a(n) _____.
5. A ligamentous connection such as between the bones of the lower leg is termed _____.
6. Movement towards the midline of the body is termed _____.
7. The special movement of the thumb that allows it to grasp an object and hold onto it is called the _____.
8. A twisting motion of the foot that turns the sole inward is termed _____.
9. The elbow joint is an example of a(n) _____ joint.
10. The joint between the carpus and metacarpus 1 of the thumb is an example of a(n) _____.
11. Which type of joints is found between the carpals? _____
12. Which ligament spans the gap between the coracoid process and the acromion? _____
13. The joints between vertebrae are example of _____ joints.
14. A _____ is a synarthrosis that binds the teeth to the bony sockets in the maxillary and the mandible.
15. An extension past the anatomical position is known as _____.
16. Spreading your fingers apart is a form of _____.
17. Nodding your head up and down is an example of _____.
18. The movement associated with chewing food is _____.
19. The pubic symphysis represents a(n) _____ articulation.
20. Small pockets of synovial fluid that reduce friction and act as a shock absorber when ligaments and tendons rub against other tissues are called _____.
21. The movement of a body part forward in a horizontal plane is called _____.
22. The movement of a body part backward in a horizontal plane is called _____.

23. The movement of a body part downward is called _____.
24. The hip bones articulate with the sacrum at the _____ joint.
25. Eversion turns the sole of the foot _____.
26. A slipped disk is the result of deterioration of the n_____
     p_____.
27. The type of movement that increases the angle between body parts is

     _____.
28. Ligaments join _____.
29. Tendons join _____.
30   All of the rheumatic diseases that affect the synovial joints are known as

     _____.

Chapter 9
Muscle Tissue

Muscle tissue is one of the four primary types of tissue. It consists chiefly of muscle cells that are highly specialized for contraction. Three types of muscle tissue exist: (1) skeletal muscle, (2) cardiac muscle, and (3) smooth muscle.

9.1 Functions of Skeletal Muscles

9.1.1 Functions

1. Produce skeletal movement
2. Maintain posture and body position
3. Support soft tissues
4. Guard entrance and exits
5. Maintain body temperature
6. Communication (speaking, writing, gesturing, and creating facial expressions)
7. Respiration—skeletal muscles of the thorax provide movements necessary for respiration
8. Storage of nutrients reserve
9. Voluntary; contracts rapidly

9.1.2 Properties

1. **Contractility**: ability to shorten forcefully

2. **Excitability**: capacity to respond to a stimulus

3. **Extensibility**: ability to be stretched beyond its normal resting length and still able be to contract

4. **Elasticity**: ability to recoil to its original resting length after it has been stretched

9.1.3 Histology

- Striated—unbranched fibers
- Several muscle cells are fused together (referred to as syncytium).
- The nuclei are peripherally located
- Gap junctions are absent between cells.
- Sarcomeres and T tubules are present.
- Two thin filaments: one thick filament

9.1.4 Physiology

- Calcium is stored and released from the **sarcoplasmic reticulum.**
- Contraction is regulated by the interaction of **troponin** and **tropomyosin.**
- Characterized by short refractory period

9.2 Types of Muscle Fibers

9.2.1 (1) **Type I Slow** Oxidative Skeletal Fibers

- They have slow contraction rates.
- They have high myoglobin content.
- They contain many mitochondria.
- Myosin ATPase activity is slow.
- ATP synthesis occurs via aerobic respiration.
- They are highly vascular.

- They have low glycogen content.
- They are fatigue resistant.
- They are suitable for endurance-type activities, such as running a marathon.

9.2.2 (2) **Type II Fast** Glycolytic Skeletal Fibers
- Low myoglobin
- Few mitochondria
- Less vascular
- High glycogen content
- Anaerobic glycolysis
- Fatigue easily
- Large-diameter fibers
- Strong, rapid contractions
- Found in arm muscles

9.3 Organization of Connective Tissue

Three layers of connective tissue are part of each muscle: (1) an epimysium, (2) a perimysium, and (3) an endomysium.

9.3.1 Epimysium
- surrounds entire muscle
- consists of a dense layer of collagen fibers
- is connected to the deep fascia, dense connective tissue layer

9.3.2 Perimysium
- surrounds each bundle of muscle fibers (fascicle)
- contains collagen and elastic fibers
- blood vessels and nerves

9.3.3 Endomysium
- surrounds individual muscle fiber within the fascicle
- is mostly elastic connective tissue
- contains a capillary network
- consists of satellite cells

9.3.4 A **tendon/aponeurosis** is a bundle of collagen fibers of epimysium, perimysium, and endomysium that attaches muscles to bone.

9.3.5 The **sarcolemma** is the cell membrane of a muscle fiber that surrounds the sarcoplasm. The sarcolemma has a characteristic transmembrane potential that renders the outside of the membrane positive in relation to the inside. A sudden change in this potential initiates the process of muscle contraction.

9.3.6 Transverse tubules are narrow tubes that are continuous with the sarcolemma and extend into the sarcoplasm. They form passageways for electrical impulses that trigger muscle fiber contraction.

9.4 Myofibrils

Myofibrils are cylindrical structures that are found inside muscle fibers. They are made up of bundles of **myofilaments** composed of **actin** (thin filaments) and **myosin** (thick filaments).

9.5 The Sarcoplasmic Reticulum

The sarcoplasmic reticulum (SR) is a membrane complex similar to the endoplasmic reticulum in a typical cell. SRs form a network around each myofibril; there they fuse with transverse tubules to form expanded chambers called **terminal cisternae** where calcium ions are stored. A pair of terminal cisternae plus a transverse tubule forms a structure called **a triad**.

9.6 Sarcomeres

The myofilaments (thin and thick) are organized into functional units called sarcomeres. A typical myofibril contains up to 10,000 sarcomeres that are arranged end to end. Sarcomeres are the smallest functional unit of a muscle fiber. Interactions between actin and myosin within these sarcomeres are responsible for muscle contraction.

9.7 The Contraction of Skeletal Muscle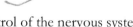

Skeletal muscle fibers contract only under the control of the nervous system. Communication between the nervous system and the skeletal muscle occurs at specialized intercellular junctions known as a **neuromuscular junction (NMJ)**.

9.7.1 A **neuromuscular (myoneural) junction** is a synapse between a motor unit and a skeletal muscle fiber. Its components are
1. Motor neuron: axon terminals
2. Neurotransmitter: acetylcholine stored in synaptic vesicles
3. Synaptic cleft
4. Skeletal muscle fiber: motor end plate

9.8 Events at a Neuromuscular Junction

Nerve impulse reaches axon terminal.
Acetylcholine is released by exocytosis.
AcH binds to receptors on the motor end plate.
Muscle action potential develops.
This causes the muscle to contact.

9.8.1 Sliding Filament Hypothesis

When a skeletal muscle fiber contracts, (1) the **H zones** and **I bands** get smaller, (2) the **zones of overlap** get larger, (3) the **Z lines** move closer together, and (4) the width of the **A band** remains constant.

9.8.2 Description

Myosin crossbridges pull thin filaments.→Overlapping occurs (filament length remain the same).→Sarcomere, muscle fiber, and muscle shorten.→Muscle contracts.

9.8.3 Sequence

Neuromuscular junction

Nerve impulse travels to axon bulbs→Acetylcholine is released and binds to receptors.→Sarcomere allows Na in and K out.→Muscle action potential develops.

9.8.4 Fiber and Filament activities

Muscle action potential opens calcium channels.→ Calcium enters the sarcoplasm and bathes the myofilaments.→ Calcium binds to troponin, which causes it to change shape.→ Troponin then pulls tropomyosin, thereby exposing myosin binding sites on actin.

9.8.5 Power stroke

ATP is bound to myosin.→ Atpase on myosin head is activated by muscle action potential.→ Energy is transferred from ATP to myosin head.→ This activates myosin cross-bridges.→ Myosin binding sites open, and myosin binds to actin.→ Myosin swivels, and ADP is released.→ The cross-bridges pull thin filaments. (This process continues until calcium ions are depleted.)

9.8.6 Relaxation

Nerve action potential ceases.→ Acetylcholinesterase (AchE) digests the excess acetylcholine.→Muscle action potential ceases.→Calcium is removed from sarcoplasm by active transport (calsequestrin). →'Troponin shape is restored.

Rigor mortis results after death due to a lack of ATP to split myosin-actin cross-bridges. This causes the muscle to become rigid.

Tension Production

The **all-or-none principle** states that individual muscle fibers contract to their fullest extent; they do not partially contract.

A **twitch contraction** is a brief contraction of all the muscle fibers in a motor unit in response to a single action potential. The process can be recorded (myogram). The myogram includes three periods: latent, contraction, and relaxation. Twitches vary in duration from 7.5 msec in eye muscle fibers to 100 msec in soleus of calf muscles.

Refractory period refers to the time when a muscle has temporarily lost excitability.

Treppe is an increase in peak tension with each successive stimulus delivered shortly after the completion of the relaxation phase of the preceding twitch. It results from a gradual increase in the concentration of calcium ions in the sarcoplasm.

Wave summation occurs when successive stimuli arrive before the relaxation phase has been completed.

Incomplete tetanus occurs if the stimulus frequency increases further. Tension production rises to a peak, and the periods of relaxation are very brief.

Complete tetanus occurs when the stimulus frequency is so high that the relaxation phase is eliminated; tension plateaus at maximal levels.

Recruitment is the process of increasing the number of active motor units. It prevents fatigue and helps provide smooth muscular contraction.

Muscle tone is a state of sustained partial contraction of portions of a relaxed muscle. This is necessary for maintaining posture.

Hypotonia leads to flaccid muscle due to decreased or lost muscle tone.

Hypertonia refers to increased muscle tone leading to rigid or stiff muscles.

Isotonic contractions occur when a muscle changes in length and moves a load. Once sufficient tension has developed to lift the load, the tension remains relatively constant through the rest of the contractile period. There are two types of isotonic contractions:

1. **Concentric contraction** occurs when the muscle tension exceeds the resistance and the muscle shortens.

2. **Eccentric contraction** occurs when the peak tension developed is less than the resistance, and the muscle elongates due to the contraction of another muscle or

the pull of gravity.

9.9 Types of Skeletal Muscle fibers

The human body has three major types of skeletal muscle fibers: fast fibers, slow fibers, and intermediate fibers.

9.9.1 Fast fibers

- Most of the skeletal muscle fibers in the body
- Can contract in 0.01 sec after stimulation
- Large diameters
- Densely packed myofibrils
- Large glycogen reserves
- Few mitochondria
- Tension directly proportional to number of myofibrils
- Fatigue rapidly

9.9.2 Slow fibers

- Small diameter
- Low contraction speed
- High fatigue resistance
- Red in color
- High myoglobin content ● Extensive capillary supply ● Many mitochondria ● Low glycolytic enzyme concentration in sarcoplasm

9.9.3 Intermediate fiber diameter

- Produce intermediate tension
- Fast contraction speed
- Intermediate fatigue resistance
- Pink in color
- Low myoglobin content
- Moderate capillary supply
- Moderate amount of mitochondria
- High glycolytic enzyme concentration in sarcoplasm

9.10 Cardiac Muscle Tissue

Cardiac muscle cells contain organized myofibrils, and the presence of many aligned sarcomeres gives the cells a striated appearance. Cardiac muscle is voluntary.

Histology	Physiology
striated	contracts without neural stimulation (automaticity)
contain single cells	Troponin and tropomyosin regulate contraction.
centrally located nucleus	Long refractory period (10 times as long as that of skeletal muscle fibers)
Gap junctions are present.	Cannot undergo wave summation nor tetanic contractions
intercalated discs present, permitting action movement	
Short, broad T tubules encircle the sarcomere at the Z lines.	

high myoglobin content	
contains many mitochondria	
highly vascular	

9.11 Smooth Muscle Tissue

Smooth muscle tissue forms sheets, bundles, or sheaths around other tissues in almost every organ. Smooth muscles around blood vessels regulate blood flow through vital organs.

- Locations: walls of hollow organs, blood vessel walls, iris, and arrector pili muscles

Histology	Physiology
*nonstriated	1. Calcium released from sarcoplasm and extracellular fluid
*spindle shaped single cells	2. Calcium binds to calmodulin and activates it.
	3. Activated calmodulin activates the kinase enzyme.
*centrally located nucleus	4. Activated kinase catalyzes transfer of PO4 from ATP to myosin cross-bridges.
*no sarcomeres or T tubules	5. Phosphorylated cross-bridges interact with actin of the thin filaments, producing shortening.
*Gap junction present in visceral muscle but absent in multi-unit muscle	6. Relaxation occurs when intracellular Ca+ levels drop.
*contains 10-15 thin filaments: 1 thick filament	
*Intermediate filaments are attached to dense bodies.	

Regeneration potential of smooth muscle is greater than other muscle types.
Types
Visceral (single-unit)
- Walls of hollow organs
- Form large networks with gap junctions
Multi-unit
- Large artery walls, bronchioles
- Few gap junctions
- Individual fibers act separately
Contraction stimuli
 Hormones, nerves, wall stretching, changes in pH, gas levels, temperature, and ion concentration

9.12 Energy Sources

The energy required to produce ATP comes from three sources: (1) creatine phosphate, (2) anaerobic respiration, and (3) aerobic respiration.

9.12.1 Creatine Phosphate

Energy from aerobic respiration is used to synthesize creatine phosphate during resting conditions.

ADP + Creatine phosphate → Creatine + ATP (Creatine kinase-enzyme)

The process occurs rapidly and is able to maintain ATP levels as long as creatine phosphate is available in the cell. It can sustain a maximum contraction for about 8-10 seconds.

9.12.2 Anaerobic Respiration does not require oxygen.
- It results in the breakdown of glucose to yield ATP and lactic acid.
- A net production of two ATP molecules and two molecules of lactic acid occurs for each molecule of glucose metabolized.

9.12.3 Aerobic Respiration
- Requires oxygen to break down glucose to produce ATP, carbon dioxide, and water
- It is more efficient than anaerobic respiration.
- Can produce up to 38 ATP molecules per glucose molecule
- Glucose + 6 O_2 + 38 ADP + 38 P → 6 CO_2 + 6 H_2O + about 38 ATP
- The rate of ATP production is slower than anaerobic respiration.
- Long-distance runners depend primarily on aerobic respiration for ATP synthesis

Oxygen Deficit and Recovery Oxygen Consumption

Oxygen deficit is the extra amount of oxygen that must be taken in by the body for restorative processes. It represents the difference between the amount of oxygen needed for totally aerobic muscle activity and the amount actually used.

Example: You use 10 L of Oxygen for a 16-second dash, but the actual amount that could be delivered to your muscles is about 2.8 L. This means that you owe 8.2 L of oxygen that must be repaid by rapid deep breathing

Review Questions

1. The dense layer of collagen fibers that surround the entire skeletal muscle is the _____.

2. Nerves and blood vessels that serve the muscle fibers are located in the connective tissues of the _____.

3. The delicate connective tissue that surrounds the skeletal muscle fibers and ties adjacent muscle fibers together is called the _____.

4. The bundle of collagen fibers at the end of a skeletal muscle that attaches the muscle to bone is called a(n) _____.

5. The advantage of a skeletal muscle fiber having many nuclei is _____

_____

_____.

6. Skeletal muscle fibers are formed from embryonic cells called _____.

7. The cell membrane of skeletal muscle is called _____.

8. The cytoplasm of a skeletal muscle fiber is called _____.

9. The skeletal muscle complex known as the triad consists of

_____.

10. At rest, the tropomyosin molecule is held in place by _____.

11. Each skeletal muscle fiber is controlled by a neuron at a single _____.
12. The space between the neuron and the muscle is the _____.
13. Receptors for acetylcholine are located on the _____.
14. Action potential are conducted into a skeletal muscle fiber by _____.
15. Define wave summation.

    _____

16. Define incomplete tetanus.

    _____

17. If a second stimulus arrives before the relaxation phase, a second, more powerful contraction occurs. This is called _____.
18. A muscle that is stimulated so frequently that the relaxation phase is completely eliminated is said to exhibit _____.
19. The increase in muscle tension that is produced by increasing the number of active motor unit is called _____.
20. The type of contraction in which the muscle fibers produce increased tension but do not shorten is called _____.
21. A resting muscle generates most of its ATP by _____.
22. What is the function of creatine phosphate?

    _____

23. When energy reserves in a muscle are exhausted and lactic acid levels increase, _____ occurs.
24. Why does the body's need for oxygen increase during the recovery period?

    _____

    _____

25. List two main characteristics of fast fibers (1) _____ (2) _____
26. The type of muscle fiber that is most resistant to fatigue is the _____ fiber.
27. Where is most of the energy required for aerobic endurance activities produced?

    _____

28. When acetylcholine binds to receptors at the motor end plate, the muscle membrane becomes _____.
29. How would blocking the activity of acetylcholinesterase affect skeletal muscle?

    _____

    _____

30. Define the term "rigor mortis".

    _____

Chapter 10
Muscular System

10.1 How a muscle produces movement

- Muscles pull tendons, and tendons pull bones or other structures.
- Muscles always cross a joint
- **Muscle origin (head):** The place where the fixed end attaches to bone or cartilage
- **Muscle insertion:** The site where the movable end attaches to another structure
- **Muscle belly:** Largest (fleshy) part of the muscle between tendons

10.1.1 Lever System

Depending on the relative position of the three elements—effort (tension), fulcrum (pivot), and load (resistance)—a lever belongs to one of three classes.

Load = R (opposes movement), Fulcrum = F (joint), Effort = E (achieves movement)

First Class: the effort is applied at one end of the lever and the load is at the other, with the fulcrum somewhere between.

- E F R
- Seesaw and scissors are examples.
- Action of the triceps muscle extending the forearm against resistance

Second class: The effort is applied at one end of the lever and the fulcrum is located at the other, with the load between them

- F R E.
- Wheelbarrow is an example.
- Act of standing on your toe, joints in the ball of the foot (fulcrum), entire body weight (load)

Third Class: The effort is applied between load and fulcrum.

- F E R
- Most common in the body
- Tweezers or forceps are examples.
- Muscle is inserted very close to the joint across which movement occurs.
- Flexing forearm on arm
- Adducting thigh

10.2 Fascicular Arrangements

All skeletal muscles consist of fascicles with varied arrangements. They are as follows:

10.2.1 Parallel

- The long axes of the fascicles run parallel to the long axis of the muscle.
- They have flat tendons
- An example is the geniohyoid muscle.

10.2.2 Fusiform

- Nearly parallel long axis; spindle-shaped muscle
- Taper toward flat tendons
- Biceps brachii and flexor pollicis brevis are examples.

10.2.3 Pennate
- The fascicles are short compared to muscle length.
- Tendon runs length of the muscle.

Types
1. **Unipennate**: Fascicles insert into only one side of the tendon—e.g., extensor digitorum longus.
2. **Bipennate**: Fascicles insert into the tendon from opposite sides. The muscle's "grain" resembles a feather. Rectus femoris is an example.
3. **Multipennate**: Several tendons attach at angles. It looks like many feathers arranged side by side. Deltoid muscle is an example.

10.2.4 Circular
- Fascicles are arranged in concentric rings around an opening.
- They form sphincters.
- Orbicularis oculi is an example.

10.2.5 Convergent
- The muscle has broad origin, and its fascicles converge toward a single tendon of insertion.
- They are triangular shaped.
- Pectoralis major muscle is an example.

10.3 Group Actions
Muscles can be classified into four functional groups as follows:
1. **Prime mover (agonist)** is a muscle whose contraction is chiefly responsible for producing a particular movement.
2. **Antagonist is a muscle whose action opposes that of a particular agonist. The triceps brachii muscle is an antagonist of the biceps brachii muscle.**
3. **Synergists**: muscles that help the prime movers by adding a little extra force to the same movement or by reducing unwanted movement
4. **Fixators**: muscles that immobilize a bone or a muscle's origin

10.4 Naming Skeletal Muscles
1. **Location of the muscle**: Temporalis muscle overlies the temporal bone; intercostalis muscles run between the ribs; radialis, femoris, and medialis, etc.
2. **Shape of the muscle:** deltoid (triangular), trapezius (trapezoid)
3. **Size of the muscle:** maximus (largest), minimus (smallest), longus (long), brevis (short)
4. **Fiber direction:** Tranversus, oblique, rectu
5. **Number of origins**: biceps, triceps, and quadriceps (2, 3, and 4 origins)
6. **Origin and insertion:** The sternocleidomastoid muscle of the neck has a dual origin on the sternum (sterno) and clavicle (cleido). It inserts on the mastoid process of the temporal bone. The origin is always named first.
7. **Action:** flexor, extensor, or adductor

10.5 Muscles of Facial Expression
- Confined primarily to head and neck

- Examples:
    1. **Frontalis:** pulls the scalp forward, raises eyebrows, wrinkles forehead
    2. **Occipitalis:** pulls the scalp backward
    3. **Orbicularis oculi:** closes eye
    4. **Orbicularis oris** constricts the opening of the lips; kissing and whistling muscle
    5. **Buccinator** muscle helps move food back across the teeth from the vestibule, provides suction for suckling at the breast.
    6. **Platysma:** lowers mandible, pulls lower lip back and down
    7. **Mentalis muscle protrudes lower lip**; wrinkles chin
    8. **Zygomaticus muscle raises lateral corners of mouth upward (smiling muscle)**

10.6 Muscle that Move the Mandible
- Mastication: chewing. Involves elevation of the mandible (**masseter, temporalis,** and **medial pterygoid** muscles) and depression of the mandible (**lateral pterygoid, digastric, mylohyoid,** and **geniohyoid** muscles)

10.7 Muscles that Move the Tongue
- Tongue movement is important in speech and swallowing
- Muscles that depress tongue: **genioglossus** and **hyoglossus**
- Sticks tongue out: **genioglossus**
- **Retracts (and elevates) tongue: styloglossus**
- Pulls down sides of tongue: **hyoglossus**
- Raises tongue: **styloglossus** and **palatoglossus**
- Pulls soft palate down: **palatoglossus**

10.8 Muscles that Move the Larynx
SUPRAHYOID MUSCLES: They anchor the tongue, elevate hyoid, and move larynx superiorly during swallowing. They are digastic, mylohyid, and geniohyoid muscles.
INRAHYOID MUSCLES: They depress the hyoid bone and larynx during swallowing and speaking. They are sternothyroid, omohyoid, and thyrohyoid muscles.

10.9 Muscles that Move the Eyeballs
10.9.1 Rectus muscles
- **Superior rectus:** moves eyeball superiority
- **Inferior rectus:** moves eyeball inferiority
- **Media rectus:** moves eyeball medially
- **Lateral rectus:** moves eyeball laterally
- **Oblique muscles:** insert onto the collateralize margin of the eyeball
- **Inferior oblique:** moves the eyeball side to side
- **Superior oblique:** moves the eyeball down and to side

10.10 Muscles that Move the Head
10.10.1 Sternocleidomastoid
- Origin: **manubrium of sternum** and **clavicle**

- Insertion: mastoid process of temporal bone and superior nuchal line of occipital bone
Action:
  - Flexes the head, rotates head to same side

10.10.2 Semispinalis capitis, splenius capitis, and **longissimus capitis:** Extend head and rotate head to the opposite side.

10.11 Muscles of the Abdominal Wall

- **Rectus abdominis, external oblique**, and **transversus abdominis:** They compress the abdomen.
- **External oblique, internal oblique**, and **quadratus lumborum:** They bend or rotate the vertebral column.
- **Rectus abdominis:** It flexes the vertebral column.
- **Quadratus lumborum:** It assists in forced respiration.

10.12 Muscles Used in Breathing

- **Diaphragm:** It changes the vertical length of the thoracic cavity.
- **External intercostalis:** It increases the lateral width of the thoracic cavity. It is a synergist of the diaphragm.
- **Internal intercostalis:** It decreases the lateral width of the thoracic cavity by drawing ribs together and depressing the rib cage. It is an antagonist to external intercostalis.
- **Quadratus lumborum, serratus anterior**, and **pectoralis minor** and **scalene group:** They maintain maximal vertical length of thoracic cavity during forced respiration.

10.13 Muscles of Pelvic Floor and Perineum

- **Levator ani** and **coccygeus:** components of the pelvic floor
- **Bulbocavernosus, deep transverse perineus**, and **urethral sphincter:** help expel last drops of urine and semen
- **Bulbocavernosus:** assists in erection, decreases diameter of vaginal orifice
- **Ischiocavernosus:** maintains erection
- **External anal sphincter:** keeps anal orifice closed

10.14 Scapular Movements

- Muscles that attach arm to thorax: **pectoralis major, latissimus dorsi**
- Deltoid and pectoralis major both act as flexors and extensors of the shoulder
- Deltoid abducts and medially and laterally rotates arm

10.15 Rotator Cuff

- They are the primary muscles holding humerus in the glenoid cavity.
- They form a cuff over the proximal humerus.
- They are involved in:
  1. Flexion (pectoralis major, deltoid, and coracobrachialis)
  2. Extension (latissimus dorsi, deltoid, teres minor, and teres major)
  3. Adduction (pectoralis major, infraspinatus, teres minor, teres major)
  4. Abduction (deltoid and suprspinatus)

     5.   Rotation (pectoralis major, latissimus dorsi, subscapularis, and teres major)

10.16 Muscles that Move the Forearm
Movement of the elbow
1. **Flexors**: biceps brachii, brachialis, and brachioradialis
2. **Extensors**: triceps brachii and anconeus
3. **Pronators:** pronator teres and pronator quadrates
4. **Supinator**: supinator and bicep brachii

10.17 Muscles that Move the Wrist, Hands and Fingers
- The insertions of these muscles are anchored by strong ligaments called flexor and extensor retinacula

Anterior group
- **Pronator teres**: pronates forearm; weak flexor of elbow. Origin is on medial epicondyle of the humerus.
- **Flexor carpi radialis**: powerful flexor of the wrist; abducts hand. Origin is on medial epicondyle of the humerus.
- **Palmaris longus**: weak wrist flexor; tenses skin and fascia of palm during hand movements. Origin is on medial epicondyle of the humerus.
- **Flexor carpi ulnaris**: powerful flexor of wrist. Origin is on medial epicondyle of the humerus.
- **Flexor digitotorum superficialis**: flexes wrist and the phalanges of fingers 2–5. Origin is on the medial epicondyle of the humerus, coronoid process of ulna, and shaft of radius.
- **Flexor polilicis longus (deep)**: flexes distal phalanx of thumb. Origin is on anterior surface of radius and interosseous membrane.
- **Flexor digitorum profundus (deep)**: slow-acting flexor of any or all fingers. Origin is on coronoid process of the ulna.
- **Pronator quadratus (deep):** prime mover of forearm pronation; helps hold ulna and radius together. Origin is at the distal portion of the ulna shaft.

10.18 Posterior Group
- **Extensor carpi radialis lingus:** extends wrist in conjunction with extensor carpi radialis and extensor carpi ulnaris
- **Extensor carpi radialis**: extends and abducts the wrist
- **Extensor carpi ulnaris**: extends wrist
- **Supinator (deep)**: assists biceps brachii to forcibly supinate forearm
- **Abductor pollicis longus**: abducts and extends thumb; origin at the surface of ulnar and radius
- **Extensor pollicis brevis** and **longus:** extend thumb

10.19 Midpalmar Muscles
- Lumbricals: flex fingers at metacarpophalangeal joints but extend fingers at interphalangeal joints
- Palmar interossei: adductors of fingers toward middle finger
- Dorsal interossei: abduct fingers toward middle finger

10.20 Thigh Movement

- Originate on coxal bone; insert onto femur
- Anterior, posterolateral (deep)
  - Anterior: flex hip: Iliacus and psoas major (Iliopsoas)
  - Posterolateral: gluteals and tensor fasciae. Extension of thigh
  - Deep: thigh rotators

10.21 Leg Movement
- **Quadriceps femoris**: anterior surface of thigh
  - Extension of the leg at the knee
  - Rectus femoris also flexes the hip
- **Sertorius**: flexes hip and knee; laterally rotates thigh
- Medial thigh muscles: adduction
- Posterior thigh muscles: hamstrings (biceps femoris, semimembranosus, and semitendinosus) Flexion and rotation of the knee
  - Popliteus muscle performs medial rotation of tibia (or lateral rotation of femur
  - Extensors of knees are rectus femoris, vastus intermedius, vastus lateralis, and vastus medialis

10.22 Extrinsic Muscles That Move the Foot and Toes
1. Tibialis anterior causes flexion at ankle; inversion of foot
2. Gastrocnemius causes extension at ankle (plantar flexion); inversion of foot; flexion of knee.
3. Soleus muscle causes extension at ankle (plantar flexion)
4. Fibularis longus causes eversion of foot and extension (plantar flexion) at ankle.
5. Flexor digitorum longus flexes the toes at joints of toes 2-5
6. Flexor hallucis longus flexes the toes at joints of great toes.
7. Extensor digitoum longus causes extension of joints of toes 2-5
8. Extensor hallucis longus causes extension at joints of great toe.

Review Questions

1. Muscles that act as valves to open and close openings are _____ muscles.
2. The paralysis of iliacus, psoas, and quadriceps femoris will affect an individual's ability to _____.
3. Skeletal muscles in which the fascicles are arranged to form a common angle with the tendon are _____ muscles.
4. A muscle whose name ends in the suffix *-glossus* would be found attached to the _____.
5. Muscles ending in the suffix *-costal* would be found in the _____.
6. The "kissing muscle" that purses the lips is the _____.
7. The origin of the frontalis muscle is the _____.
8. The origin of the occipitalis is the _____
9. The auricularis muscle acts to _____.
10. The mentalis muscle inserts on the _____.
11. What is the action of corrugator supercilii? _____
12. The muscle that inserts on the coronoid process of the mandible is the _____.
13. The muscle that elevates the tongue is the _____.
14. Name the muscles of the rotator cuff. (1) _____ (2) _____ (3) _____ (4) _____

15. Muscle fibers in skeletal muscle form bundles called _____.

16. The end of a muscle that remains stationary when the muscle contracts is called the _____.

17. The end of a muscle that is attached to the point that moves when the muscle contracts called _____.

18. The _____ allows you to look down.

19. The _____ allows you to look up.

20. The _____ muscle is the strongest jaw muscle.

Match the following terms with their main action.

21.	abductors	turn the hand palm posteriorly
22.	adductors	increase the angle of a joint
23.	extensors	move the bone away from midline
24.	flexors	decrease the angle of a joint
25.	pronators	turn the hand palm anteriorly
26.	supinators	move the part toward the midline

Match the following muscles with their location.

27.	Neck	gluteus maximus
28.	Back	biceps brachii
29.	Chest	trapezius
30.	abdominal wall	pectorialis major
31.	Shoulder	adductor magnus
32.	upper arm	semitendinosus
33.	Forearm	external oblique
34.	Buttocks	deltoid
35.	Thigh	sternocleidomastoid
36.	Leg	vastus lateralis

Chapter 11
The Nervous Tissue

The nervous system includes all the neural tissue in the body. The basic functional unit of the nervous system is an individual cell called a neuron. Supporting cells or neuroglia separate and protect the neurons, provide a supportive network for neural tissue, act as phagocytes, and help to regulate the composition of the interstitial fluid.

11.1 Functions of the Nervous System

The nervous system plays a key role in maintaining homeostasis by performing three vital functions.
1. Sensory function: Changes in internal and external environment are detected and transmitted by sensory structures called receptors to the central nervous system.
2. Integrative functions: It analyzes the sensory input and decides the next course of action.
3. Motor functions: It responds to the changes by stimulating the effector organs, such as muscle to contract or glands to secrete their products.

11.2 Nervous System Divisions
1. The central nervous system (CNS) consists of the brain and spinal cord. The CNS is responsible for integrating, processing, and initiating responses to sensory data and motor inputs.
2. The peripheral nervous system (PNS) consists of cranial and spinal nerves. It has sensory (efferent) and motor (afferent) components.
   a. The sensory system includes a variety of different receptors as well as sensory neurons.
   b. The motor system conducts nerve impulses from the CNS to muscles and glands.

3. The PNS is also subdivided into somatic (voluntary) and autonomic (involuntary) nervous systems.
   a. The somatic nervous system (SNS) consists of neurons that conduct impulses from cutaneous and special sense receptors to the CNS and motor neurons that conduct impulses from the CNS to the skeletal muscle tissue.
   b. The autonomic nervous system (ANS) contains sensory neurons from the visceral organs and motor neurons that convey impulses from the CNS to smooth muscle tissue, cardiac muscle tissue, and glands. The ANS includes a sympathetic division and a parasympathetic division.

11.3 Histology of the Nervous Tissue

11.3.1 Neuroglia (glial cells) help to maintain homeostasis of fluids around the neurons. Unlike neurons, they have mitotic potential.

Types of Neuroglia
1. CNS: **astrocytes, oligodendrocytes, microglia,** and **ependymal cells**
2. PNS: **neurolemmocytes (Schwann cells)** and **satellite cells.**

11.3.2 Neuroglia of the Central Nervous System
1. **Ependymal cells** contain a single layer of epithelium. They line the ventricles of brain and spinal cord, where they help to circulate cerebrospinal fluid (CSF)
2. **Astrocytes** are star-shaped cells with many processes. They are located in the gray matter, where they are referred to as protoplasmic astrocytes, and in the white matter (fibrous

astrocytes). The functions of astrocytes include:
  i. Neurotransmitter metabolism
  ii. K balance for nerve impulse
  iii. Neuron migration
  iv. Blood-brain barrier
  v. Support network for neurons
  vi. Link neurons to blood vessels

3. **Oligodendrocytes** are smaller than atrocytes and have fewer processes. They have round or oval cell bodies. Groups of oligodendrocytes make up the myelin sheath along the length of an axon.
4. **Microglia** are small, ovoid cells with relatively long processes. They are capable of migrating through neural tissue, where they engulf pathogens by phagocytosis.

11.3.3 Neuroglia of the Peripheral Nervous System
1. **Satellite cells** surround neuron cell bodies within ganglia.
2. **Schwann cells** (neurolemmocytes) surround and form myelin sheaths around the larger nerve fibers in the PNS.

11.3.4 Neurons
Most neurons consist of a cell body (soma), many dendrites, and usually a single axon.
- The dendrites conduct impulses from receptors or other neurons to the cell body.
- The axon conducts nerve impulses from the neuron to the dendrites or cell body of another neuron or to an effector organ such as muscle or gland.
- A nerve fiber refers to a neuron process—dendrite or axon. The processes of a neuron are arranged into nerves in the PNS and tracts in the CNS. Nerve cell bodies in the PNS form clusters called ganglia. Cell bodies in the CNS are called nuclei.

11.3.5 Classification of Neurons
Neurons are also categorized by function as (1) sensory neurons, (2) motor neurons, or (3) interneurons.
1. **Sensory neurons (afferent)**: The sensory neurons deliver information from sensory receptors to the CNS. They are unipolar neurons.
   Types
     i. Somatic sensory neurons: Monitor outside world and our position within it
     ii. Visceral sensory neurons: Monitor internal conditions
2. **Motor neurons** (efferent) carry instructions from the CNS to tissues, organs, or systems.
   Types
     i. Somatic motor neurons (skeletal muscles)
     ii. Visceral motor neurons (innervate all peripheral effectors other than skeletal muscle)
3. **Interneurons** are mostly found within the brain and spinal cord, although some are found in the autonomic ganglia. They function in the distribution of sensory information and coordination of motor activities.

11.4 Axonal transport is comprised of two types:
   1. Slow axonal transport (axoplasmic flow) involves one-way flow toward axon terminals bringing new axoplasm to the axons.
   2. Fast axonal transport involves bi-directional transport. This process is used to move organelles, viruses, and membrane materials.

11.5 Neuron Communication Factors

The passive forces acting across the membrane are both chemical and electrical in nature.

- Chemically regulated channels open and close when they bind to specific chemicals. These are common on the dendrites and cell body of a neuron.
- Voltage-regulated channels are found in excitable membranes such as the sarcolemma of muscle fibers and cardiac muscle cells. Examples are sodium channels, potassium channels, and calcium channels.

11.6 Resting Membrane Potential (RMP)

The potential difference or voltage in a resting neuron is called resting membrane potential; if a potential difference exists, the membrane is said to be polarized. This is due to differential charges across the cell membrane. The outside is more positive and the inside is more negative. The switching of ion channels allows current to flow. The resting membrane potentials depend upon the following factors:

1. Extracellular fluid (ECF) contains sodium and chloride.
2. Intracellular fluid (ICF) contains potassium, organic phosphate, and amino acids.
3. Membranes are moderately permeable to K and slightly permeable to Na.
4. Na/K active transport pumps help maintain voltage gradient: 3 Na out : 2 K in.

11.6.1 Types of Ion Channels

Some integral proteins contain a central pore, or channel, that forms a passageway completely across the cell membrane. The channel permits the movement of water and solutes (ions) across the cell membrane. There are two types of ion channels:

1. **Leakage channels (nongated)** are always open—more for K, fewer for Na.
2. **Gated channels** open and close in response to stimuli. There are four types of gated channels:
   i. **Voltage-gated channels** open and close in response to changes in membrane potential.
   ii. **Chemically gated channels** open when an appropriate neurotransmitter binds to a receptor on the membrane.
   iii. **Mechanically gated channels** open and close in response to mechanical stimuli such as sound and touch.
   iv. **Light-gated** channels open and close in response to light stimulus.

11.6.2 Ion Channels and Membrane Potentials

There are two types of signals produced by a change in membrane potential:

1. Action potentials involve voltage-gated channels for Na and K leading to **depolarization/ repolarization of the membrane as the channels open in sequence.** It involves the all-or-none principle and lasts for about 1 msec.
2. **Graded potentials** involve gated channels that are opened by chemical, mechanical, and light stimuli. They are short-lived and involve only local changes. Intensity varies depending on the intensity of the stimulus, the number of channels opened, and how long the channels are opened.

During the **refractory period**, another impulse cannot be generated at all (absolute refractory period). Only a **superthreshold** stimulus can trigger an impulse.

11.7 Types of Impulse Conduction

    i.   **Continuous conduction** occurs on an umyelinated nerve, causing depolarization in each area of the membrane. Impulse conduction is very slow in unmyelinated nerves.

    ii.   **Saltatory conduction** occurs in myelinated nerves. Depolarization occurs only in neurofibril nodes. The nodes have many voltage-gated sodium channels. Impulses "jump" from node to node. This is a fast and energy-efficient method of conduction.

11.8 Impulse Speed

The diameter of the axons also affects the propagation speed. An axon behaves like an electrical cable: the larger the diameter, the lower the resistance. Axons are classified into three groups according to the relationships among the diameters, myelination, and propagation speed.

Impulse speed is unrelated to impulse strength but depends upon the diameter of the fibers, temperature, and myelination.

Fiber types

    i.   **A Fibers** are mostly somatic sensory and motor fibers serving the skin, the skeletal muscles, and joints. They have the largest diameter and thick myelin sheaths. They are involved in reflexes and survival.

    ii.   **B fibers** are lightly myelinated and transmit impulses at an average rate of 15 m/s (40 mph). They connect visceral organs to the central nervous system.

    iii.   **C fibers** have the smallest diameter and are unmyelinated. They are incapable of salutatory conduction and conduct impulses at a slow rate (1 m/s) (2 mph). They make up the sensors in the skin, visceral pain receptors, and visceral efferent fibers.

11.9 Synaptic Transmission

A **synapse** is the functional junction between one neuron and another or between a neuron and an effector, such as a muscle or gland. There are two types of synapses:

    i.   **Electrical synapses** correspond to the gap junctions found between certain body cells. They spread ionic current directly from cell to cell.

    ii.   **Chemical synapses** are specialized for release of chemical neurotransmitters.

11.9.1 Transmission events

1. Impulse travels to the axonal terminal of the presynaptic neuron.
2. Voltage-gated calcium channels open.
3. Calcium enters the cell.
4. Neurotransmitter is released from the synaptic vesicle by exocytosis.
5. Polarization of the postsynaptic membrane changes. Two types of postsynaptic potential develop at neuron-to-neuron synapses:

    i.   Excitatory postsynaptic potentials (EPSP): Local graded polarization occurs to trigger action potential at the axon hillock.

    ii.   Inhibitory postsynaptic potentials (IPSP): Binding of neurotransmitters at inhibitory synapses reduces a postsynaptic neuron's ability to generate action potential. This induces hyperpolarization of the postsynaptic membrane, making it more permeable to potassium ions.

6. Neurotransmitters are removed by the enzyme acetylcholinesterase (AchE).

11.9.2 Summation by the Postsynaptic Neuron

EPSP can add together (summate) to influence the activity of the postsynaptic neuron. Two types of summation may occur.

- Spatial summation: Spatial summation involves the accumulation of neurotransmitters from two synapses that are active simultaneously, forming graded potentials. These graded potentials summate at the trigger zone to form a graded potential that exceeds the threshold that is able to stimulate an action potential.

- Temporal summation: Temporal summation involves the accumulation of neurotransmitters from one presynaptic bulb that is active repeatedly. The first action potential causes the production of a graded potential that does not reach threshold at the trigger zone. The second action potential produces a second graded potential that summates with the first to reach threshold, thereby stimulating an action potential.

11.9.3 Presynaptic Inhibition and Presynaptic Facilitation

Presynaptic facilitation increases neurotransmitter release at adjacent synapses and also increases stimulation.

Presynaptic inhibition decreases neurotransmitter release at adjacent synapses and decreases stimulation.

11.10 Neurotransmitters and Their Receptors

Neurotransmitters, along with electrical signals, are the "languages" of the nervous system—the means by which each neuron communicates with others to process and send messages to the rest of the body.

1) **Acetylcholine** is released at the neuromuscular junctions within the CNS and binds to cholinergic receptors. It is degraded to acetic acid and choline by the enzyme acetylcholinesterase (AchE).

2) **Amino acids** occur in all cells in the body and are important in many biochemical reactions. Examples include **gamma-aminobutyric acid (GABA), glycine, aspartate**, and **glutamic**. They act by opening chlorine channels.

3) **Biogenic amines** include the catecholamines such as dopamine, norepinephrine, and epinephrine.

11.11 Neuronal Circuits

The patterns of synaptic connections in neuronal pools are called circuits. There are four basic types of circuit patterns.

1) Diverging circuits: One presynaptic neuron stimulates many postsynaptic neurons. An example is when a single neuron of the brain activates several motor neurons in the spinal cord and consequently thousands of skeletal muscle fibers.

2) Converging circuits: Several presynaptic neurons stimulate one postsynaptic neuron, resulting in strong stimulation or inhibition. The sight and smell of a delicious food can trigger a flood of reactions, such as mouth watering, hunger, etc.

3) Reverberating circuits: One presynaptic neuron stimulates a series of impulses, and the impulses are recycled to cause positive feedback until one neuron in the circuit fails to fire. Examples of reverberating circuits include the sleep-wake cycle and breathing.

4) Parallel after-discharge circuits: The presynaptic neuron stimulates several neurons arranged in parallel arrays. Each member of the group stimulates a common postsynaptic neuron. Parallel after-discharge circuits are involved in complex mental processes such as solving mathematical equations.

Review Questions

1. The brain and spinal cord comprise the _____ nervous system.

2. Voluntary control of skeletal muscles is provided by the _____ nervous system.

3. The part of the peripheral nervous system that brings information to the central nervous system is _____.

4. The largest and most numerous of the glial cells in the central nervous system are the _____.

5. The myelin sheaths that surround the axons of some of the neurons in the CNS are formed by _____.

6. The type of glial cells that are especially obvious in damaged tissue in the CNS are the _____.

7. The neurilemma of axons in the peripheral nervous system is formed by _____.

8. Glial cells found surrounding the cell bodies of peripheral neurons are _____.

9. The cytoplasm surrounding the nucleus of a neuron is called the _____.

10. Aggregations of fixed and free ribosomes in neurons are referred to as _____.

11. The axon is connected to the soma by the _____.

12. Branches that sometimes occur along the length of an axon are called _____.

13. Axons terminate in a series of fine extensions known as _____.

14. Neurotransmitters are released from the _____.

15. Neurons that have one axon and one dendrite are called _____.

16. Sensory neurons of the PNS are _____.

17. At the normal resting potential of a typical neuron, its ion exchange pump transports three _____ ions for two _____ ions

18. Membrane channels that are always open are called _____ channels.

19. Opening of sodium channels in the membrane of a neuron results in _____.

20. List the main steps in the generation of an action potential in sequence. Step #1 is listed for you
    1. A graded depolarization brings an area of an excitable membrane to threshold.
    2.
    _____

    3.
    _____

    4.
    _____

5.

_____

6.

_____

7.

_____

21.   Theall-or-noneprinciplestates

_____.

22.   Which fiber type has the greatest rate of impulse conduction? _____

23.   Adrenergic synapses release the neurotransmitter _____.

24.   Cholinergic synapses release the neurotransmitter _____.

25.   All of the nervous tissue outside the central nervous system comprises the _____ nervous system.

26.   The _____ division of the nervous system brings sensory information to the central nervous system.

27.   _____ is a wrapping produced by some glial cells that contains lipids and proteins.

28.   The gaps between adjacent wrappings on an axon are called _____.

29.   The minimum amount of stimulus required to depolarize an excitable membrane and generate an action potential is known as the _____.

30.   The time during which an excitable membrane cannot respond to further stimulation regardless of the stimulus strength is the _____.

31.   When do excitatory postsynaptic potentials (EPSPs) occur?

_____

32.   What is the effect of inhibitory postsynaptic potentials (IPSPs)?

_____

33.   A temporal summation occurs when_____.

34.   A spatial summation occurs when

_____.

35.   Any stimulus that opens a(n) _____ channel will produce a graded potential.

Chapter 12

Spinal Cord and Spinal Nerves

12.1 The adult spinal cord is approximately 45 cm in length and has a maximum width of 14 mm.

- Extends from medulla oblongata to the second lumbar vertebra (L2) in adults and to L3-L4 in newborn.
- It has two enlargements: **cervical** and **lumbar**.
- Its tapered inferior end is termed the conus medullaris.
- Cauda equina is a collection of nerves roots at the inferior end of the vertebral canal.
- Filum terminale is a fibrous extension of the pia mater.

12.1.1   Cross-sectional Anatomy of the Spinal Cord

The spinal cord is somewhat flattened from front to back and contains two grooves:

- Anterior median fissure
- Posterior median sulcus (shallow)

Gray Matter

i.   It is surrounded by white matter.

ii.   It is organized into horns—a thin posterior (dorsal) horn, and a larger anterior (ventral) horn.

iii.   Small lateral horns exist in levels of the cord associated with the autonomic nervous system.

iv.   The "body" of the butterfly contains the gray and white commissure.

v.   The gray matter contains neuron cell bodies (soma), neuroglia, and unmyelinated processes of motor neurons.

vi.   Information is processed at the gray matter.

White Matter

i.   It surrounds gray matter.

ii.   It is organized into three columns (funiculi): anterior (ventral), posterior (dorsal), and lateral columns.

iii.   It consists of myelinated axons of motor and sensory neurons.

iv.   It forms fibers of ascending (sensory) and descending (motor) tracts (pathways).

v.   It conducts nerve impulses.

12.2 Spinal Cord Protection

The spinal cord is protected by bone, meninges, and cerebrospspinal fluid (CSF).

The **meninges** are three coverings that run continuously around the spinal cord and brain.

1.   The **outermost** layer is the dura mater.
2.   The middle layer is the **arachnoid**.
3.   The innermost layer is the **pia mater**, a connective tissue that adheres to the surface of the spinal cord and brain.

Cerebrospinal fluid is formed in the choroid plexus of the brain. It functions as a mechanical shock absorber. It circulates in subarachnoid space and brain ventricles.

12.3 Organization of White Matter

The white matter on each side of the spinal cord is divided into three white columns, or funiculi:

1) The **posterior white columns**: They lie between the posterior gray horns and the posterior median sulcus.
2) The **anterior white columns**: They lie between the anterior gray horns and the anterior median fissure.
3) The **lateral white column**: It is the white matter between the anterior and posterior columns on each side of the spinal cord.
4) The **anterior white commissure**: It is the region where axons cross from one side of the spinal cord to the other.

Tracts: Tracts are bundles of axons in the central nervous system that have relatively uniform diameter, are myelinated, and have high impulse conduction speed. All the axons within a tract relay the same type of information in the same direction.

- **Ascending tracts:** They carry sensory information toward the brain.
- **Descending tracts:** They covey motor commands from the brain to the spinal cord.

12.4 Spinal Nerves

Every segment of the spinal cord is connected to a pair of spinal nerves. Surrounding each spinal nerve is a series of connective tissue layers. They are as follows:

1. The **epineurium** is the outermost layer of the dense network of collagen fibers that surrounds groups of fascicles and fuses with the dura mater.
2. The **perineurium** is the middle layer; it divides the nerves into compartments (fascicles).
3. The **endoneurium** is the innermost layer that surrounds individual axons.

12.5 Peripheral Distribution of Spinal Nerves

Each spinal nerve has the following rami (branches):

1. **Dorsal rami** contains somatic motor and visceral motor fibers that innervate the skin and deep muscles of the back.
2. **Meningeal branches** innervate vertebrae and their ligaments.
3. **Rami communicantes (communicating rami)** carry axons associated with the sympathetic division of the autonomic nervous system.
4. The ventral rami are distributed in two ways:
   - Ventral rami in the thoracic region, T2–T12, form intercostal nerves, which innervate the intercostal muscles and the skin over the thorax.
   - All other rami join one another lateral to the vertebral column, forming nerve plexuses (intermingling of the nerves).

12.6 Nerve Plexuses

1. The cervical plexus (ventral plexus C1–C4) supplies the skin and muscles of head, neck, upper shoulder, and diaphragm.
   - The most important derivative of the cervical plexus is the phrenic nerve, which innervates the diaphragm.
2. The brachial plexus (ventral rami C5–C8, T1) supplies the shoulder and upper extremities. The major derivatives are as follows:
   - The axillary nerve innervates the teres minor and deltoid muscles to laterally rotate and abduct the arm, respectively.
   - Radial nerves innervate the extensor muscles of the upper limb, the supinator muscle, and the brachioradialis.

- The musculocutaneous nerve innervates the anterior muscles of the arm as well as providing cutaneous sensory innervation to part of the forearm.
- The ulnar nerve innervates several muscles of the forearm, such as the flexor carpi ulnaris, flexor digitorum profundus, and adductor pollicis, resulting in flexion and adduction movements of the wrist and fingers.
- The median nerve innervates all but one of the flexor muscles of the forearm and the muscles of the base of the thumb.

3. The lumbar plexus (ventral rami L1–L4) and sacral plexus (ventral rami L4–L5, S1–S4) supply the anterolateral abdominal wall, external genitals, and part of the lower extremities, buttocks, and perineum

The major nerve derivatives are as follows:
- The obturator nerve innervates the medial thigh through the obturator externus (rotation), adductor magnus, longus, and brevis (adduction).
- The femoral nerve innervates the iliopsoas and sartorius muscles and the quadriceps femoris group.
- The tibial nerve innervates most of the posterior thigh and leg muscle.
- The common fibular (peroneal) nerve innervates the anterior and lateral muscles of the leg and foot.
- The tibial nerve and the common (peroneal) nerves combine to form the sciatic nerve (the largest peripheral nerve in the body).

4. The coccygeal plexus is formed from the ventral rami of spinal S5 and the coccygeal nerve (C0). It provides innervation to the muscles of the pelvic floor.

Reflexes

The basic components of the reflex arc are:
1. a sensory receptor
2. a sensory neuron
3. an interneuron
4. a motor neuron
5. an effector

Reflexes produce responses to changes in the environment. They are automatic, fast, and predictable. They help the body to maintain homeostasis. Reflexes may be **spinal, cranial, somatic (skeletal muscles)**, or **autonomic** (heart rate, respiration, digestion, urination, and defecation)

Review Questions

1. **In the spinal cord, white matter is organized into ascending and descending groupings called _____.**
2. **The nerves serving the upper and lower limbs arise from _____ and _____.**
3. **The fibrous extension of the pia mater is called _____.**
4. **Blood vessels directly supplying the spinal cord are found in the _____.**
5. **The projections of gray matter toward the outer surface of the spinal cord are called**

_____.

6. Axons crossing from one side of the spinal cord to the other within the gray matter are found in the _____.

7. The white matter of the spinal cord contains

_____.

8. The outermost connective tissue covering of spinal nerves is the _____.

9. The dorsal root ganglia contain

_____.

10. The ventral root of a spinal nerve contains

_____.

11. The dorsal and ventral roots of each spinal segment unite to form a(n)

_____.

12. The preganglionic fibers that connect a spinal nerve with an autonomic ganglion in the thoracic and lumbar region of the spinal cord are the

_____.

13. The _____ of each spinal nerve provides sensory and motor innervation to the skin and muscles of the back.

14. Muscles of the neck and shoulder are innervated by spinal nerves from the _____region.

15. Spinal nerves from the sacral region of the spinal cord innervate the _____ muscles.

16. Sensory and motor innervations of the skin of the sides and front of the body are provided by the _____.

17. The joining of the ventral rami of adjacent spinal nerves is termed a(n)

_____.

18. A nerve of the cervical plexus that innervates the diaphragm is the _____.

19. The ventral rami of spinal nerves C5 to T1 form the _____ plexus.

20. The ulnar nerve is found in which plexus? _____

21. A reflex that moves the limb away from a painful stimulus is a _____ reflex.

22. A reflex that prevents a muscle from exerting too much tension is the _____ reflex.

23. The specific region of the skin that is innervated by a specific spinal nerve is called a(n)

_____.

24. A complex, interwoven network of nerves is called a(n) _____.

25. _____ reflexes involve skeletal muscles.

26. The most complicated spinal reflexes are called _____.

27. The condition in which a person loses sensation and motor control of the arms and legs is termed _____.

28. The loss of motor control of the legs is termed _____.

29. The major nerve of the cervical plexus is termed the _____.

30. What effect would the severing of the dorsal root of a spinal nerve have on the body?

_____

Chapter 13

The Brain and Cranial Nerves

A typical adult brain contains almost 98 percent of the body's neural tissue. It weighs about 1.4 Kg or 3 lb. The brain is the center for intellect, emotions, behavior, and memory.

13.1 Embryonic Development of the Brain

The neural tube's anterior end begins to expand and constrict to mark off the three primary brain vesicles: (prosencephalon, mesencephalon, and rhombencenphalon).

- The diencephalon develops into the thalamus and hypothalamus.
- The telencephalon forms the cerebrum.
- The mesencephalon develops into the midbrain.
- The myelencephalon forms the medulla.
- The metencephalon develops into the pons and cerebellum.

13.1.1 The Adult Brain

The adult brain can be studied in terms of the embryonic scheme or the medical scheme. In terms of the medical scheme, the principal parts of the adult brain are (1) brain stem, (2) diencephalon, (3) cerebrum, and (4) cerebellum.

1. Brain stem: The brain stem is continuous with the spinal cord. It regulates involuntary activities of the body. It is made up of the medulla oblongata, pons, and midbrain.
2. Diencephalon: It is continuous with the brain. It plays a role in integrating conscious and unconscious sensory information and motor commands. It consists of the epithalamus, thalamus, and hypothalamus.
3. Cerebrum: It covers the diencephalon. It contains the right and left hemispheres. It is the thinking part of the brain.
4. Cerebellum: It lies posteriorly to the brain stem. It is the coordination center of the brain.

13.2 Brain Protection

The delicate tissues of the brain are protected from mechanical forces by the following structures:

1. Cranial bones: prevent direct trauma and infections
2. Cranial meninges: prevent direct trauma and infections, also provide support
3. Cerebrospinal fluid: functions as mechanical shock absorber
4. Blood: provides temperature control
5. **Neuroglia**: form the blood-brain barrier; provide immunity

13.2.1 Cerebrospinal Fluid (CSF)

Cerebrospinal fluid (CSF) completely surrounds and bathes the exposed surfaces of the central nervous system. It has several important functions:

1. Mechanical shock absorber
2. Chemical protection
3. Circulation of nutrients, chemical messengers, and waste products

13.2.2 Formation of CSF

CSF is formed by the following structures:

1. Ependymal cells
2. The choroid plexus, which is a combination of specialized ependymal cells and capillaries

3. Walls of ventricles

13.2.3 Circulation of CSF

The choroid plexus produces CSF at a rate of about 500 ml/day. The total volume of CSF at any moment is about 150 ml. The entire CSF is replaced every eight hours. CSF travels as follows:

Lateral ventricles →Interventricular foramen → third ventricle → Cerebral Aqueduct → fourth ventricle →Median and lateral apertures → Subarachnoid space → Around brain and spinal cord → Arachnoid villi reabsorb → Dural venous sinuses
Note: Volume made = volume reabsorbed

13.3 Blood Supply to the Brain

The high-energy demands of the brain for energy and nutrients are met by the extensive circulatory supply. Arterial blood reaches the brain through the internal carotid arteries and the vertebral arteries. Most of the veins from the brain leaves the cranium in the internal jugular veins, which drain the dural sinuses.

13.3.1 Blood-brain Barrier (BBB)

Neural tissue in the CNS is isolated from the general circulation by the blood-brain barrier (BBB) formed by:
1. astrocytes
2. capillary endothelium tight junctions
3. continuous basement membranes

Astrocytes selectively transport materials. The BBB is weakest around circumventricular organs (hypothalamus, capillaries of the posterior lobe of the pituitary gland, capillaries of the pineal gland, and capillaries of the choroid plexus).

13.4 The Brain Stem

The brain stem is made up of the following parts: medulla oblongata, pons, reticlar formation, midbrain. and medial lemniscus.

13.4.1 The **Medulla oblongata** is continuous with the upper part of the spinal cord.

- It contains nuclei that are reflex centers for the heart rate, respiratory rate, vasoconstriction, swallowing, coughing, vomiting, sneezing, and hiccupping.
- It contains nuclei of origin for cranial nerves VIII (cochlear and vestibular branches) through XII.
- It contains olivary and vestibular nuclei.

13.4.2 The Pons

The pons links the cerebellum with the mesencephalon, diencephalun, cerebrum, and spinal cord. Important features and regions are as follows:
- It relays nerve impulses related to voluntary skeletal movements from the cerebral cortex to the cerebellum.
- It contains the nuclei for cranial nerves V through VII and the vestibular branch of

VIII.

- The pons also contains the pneumotaxic and apneustic areas, which help control respiration along with the respiratory center in the medulla.

13.4.3 The Midbrain or Mesencephalon

The midbrain is just superior to the pons and is associated with the nuclei of cranial nerves III, IV, and V.

Fiber tracts:

- Cerebral peduncles connect parts of the brain to each other and to the spinal cord.
- Superior cerebellar peduncles connect the midbrain to the cerebellum.
- The tectum (roof of the mesencephalon) has two pairs of sensory nuclei.
  - The corpora quadrigemina is made up of superior and inferior colliculi
  - Superior colliculi control reflex movements triggered by visual stimuli
  - Inferior colliculi control reflex movements triggered by auditory stimuli
  - The substantia nigra controls subconscious muscle activities.

13.4.4 Reticular Formation

This is a cluster of gray matter dispersed throughout the brain stem. It contributes to muscle tone. It forms the reticular activation system (RAS) that helps to maintain consciousness. It has both sensory and motor functions.

13.4.5 Medial Lemniscus

It runs throughout the brain stem. It is made up of fiber tracts. Its function is to relay sensory information from the medulla oblongata to the thalamus.

13.5 The Cerebellum

- The cerebellum occupies the inferior and posterior aspects of the cranial cavity. It consists of two hemispheres and a central, constricted vermis.
- It is attached to the brain stem by three pairs of cerebellar peduncles.
- The cerebellum functions in the coordination of skeletal muscle contractions, posture maintenance, balance, and normal muscle tone.

13.6 The Diencephalon

The diencephalon plays a vital role in integrating conscious and unconscious sensory information and motor commands. It consists of the epithalamus (which contains the pineal gland), thalamus, and hypothalamus.

13.6.1 The **pineal gland** secretes melatonin to influence diurnal cycles in conjunction with the hypothalamus.

13.6.2 The **thalamus**

- forms the lateral walls of the third ventricles
- is composed of mostly gray matter with paired oval masses and is linked to the intermediate mass
- has white matter that divides gray matter into anterior, medial, and lateral nuclear groups
- relays sensory input to cortex via the following nuclei:
  - i.   medial genticulate nuclei: hearing
  - ii.   lateral genticulate nuclei: vision
  - iii. ventral posterior nuclei: taste and general senses
- does not relay olfaction (sense of smell)

- relays motor impulses via the following nuclei:
    i. ventral lateral nuclei: voluntary actions
    ii. ventral anterior nuclei: voluntary actions and arousal
- interprets sensations of pain, touch, temperature, and proprioception (position)
- is involved in emotions

13.6.3 The **hypothalamus**
- forms the floor and inferior of the third ventricle
- has a **mammillary region** containing mammillary bodies that act as relay stations for smell reflexes
- has a **tuberal region** containing the infundibulum, which links to the pituitary gland
- has a **medial eminence** that synthesizes hormones regulating adenohypophysis
- has a **supraoptic region** that makes antidiuretic and oxytocin hormones (ADH and OT)
- has a **preoptic region** that regulates some autonomic functions

13.6.3a Functions of the Hypothalamus
- Controls and integrates autonomic activities such as heart rate, respiration, and blood pressure
- Acts as the principal intermediary between the nervous system and the endocrine system by releasing regulating hormones that stimulate or inhibit specific cells in the anterior pituitary gland
- Acts as the center of rage and aggression
- Regulates temperature
- Regulates food intake; contains feeding and satiety centers
- Contains the thirst center that regulates water intake
- Helps to coordinate wake/sleep cycles

13.7 The Cerebrum
The cerebrum is the largest part of the brain. Conscious thoughts and all intellectual functions originate from the cerebral hemispheres.

13.7.1 General Layers
The outer portion (containing gray matter) is the cerebral cortex, which contains billions of neurons. The inner portion (which contains white matter) contains fiber tracts that connect parts of the brain with itself and other parts of the nervous system.

13.7.2 Landmarks of the Cerebral Cortex
    i. Gyrus (Gyri): wrinkles, raised area
    ii. Fissures: deep, wide groove
    iii. Sulcus (sulci): shallow groove, depression
    iv. The cerebrum is nearly separated into right and left halves, called hemispheres, by the longitudinal fissure. Internally it remains connected by the corpus callosum.

13.8   Fiber Direction
- Association fibers: connect and transmit impulses between gyri in the same hemisphere (ipsilateral gyri)
- Commisural fibers: connect gyri in one cerebral hemisphere to the corresponding gyri in the opposite hemisphere (contralateral gyri)
- Projection fibers: form ascending and descending tracts that transmit impulses from the cerebrum to other parts of the brain and spinal cord.

13.9 Hemispheric Lateralization

    i.   The left hemisphere is more important for right-handed control, spoken and written language, and numerical and scientific skills.

    ii.   The right hemisphere is more important for left-handed control, musical and artistic awareness, space and pattern perception, insight, imagination, and generating mental images of sight, sound, touch, taste, and smell.

13.10 Basal Ganglia (Cerebral Nuclei)

- Functional group that comprises the caudate nucleus and lentiform nucleus
- The lentiform nucleus consists of a medial globus pallidus and a lateral putamen.
- Functions: automatic skeletal movements and muscle tone regulation

13.11 Limbic System

The limbic system is found in the cerebral hemispheres and the diencephalon.

13.11.1 Functions of Limbic System

- Emotional behavior involved with survival such as pain, pleasure, anger/rage, fear, sorrow, affection, and sexuality
- It works with hippocampus to store memories.

13.12 Motor and Sensory Areas of the Cortex

13.12.1 Sensory areas deal with sensory impulses. They consist of the following areas:

1. Primary (general) somatosensory area
2. Primary visual area: occipital lobe
3. Primary auditory area: temporal lobe
4. Primary gustatory area: anterior portion of the insula
5. Primary olfactory area: temporal lobe

13.12.2 Motor areas control motor responses. They consist of the following areas:

1. Primary motor area
2. Motor speech (Broca's area)

13.12.3 Association areas handle integrative functions. The areas involved are:

1. Somatosensory association area allows you to recognize light touch.
2. Visual association area monitors the patterns of activity in the visual cortex and interprets the results.
3. Auditory (Wernicke's) association area monitors activity in the auditory cortex, handles word recognition.
4. Gnostic area is the analytical center
5. Premotor area
6. Frontal eye field area controls learned eye movements.

13.13 Cranial Nerves

The cranial nerves are numbered using Roman numerals

I. **O**lfactory. Sensory nerves of smell
II. **O**ptic. Sensory nerve of vision
III. **O**cculomotor. It supplies four of the six extrinsic muscles that move the eye ball
IV. **T**rochlear. It innervates the pulley-shaped ligament in the orbit
V. **T**rigeminal. It supplies sensory fibers to the face and motor fibers to the chewing muscels (largest cranial nerve)
VI. **A**bducens. Controls extrinsic eye muscle that abducts the eyeball
VII. **F**acial. Innervates muscles of facial expression
VIII. **V**estbulocochlear. Sensory nerve for hearing and balance
IX. **G**lossophryngeal. Innervates tongue and pharynx
X. **V**agus. "wanderer" It extends beyond the head and neck to the thorax and abdomen.
XI. **A**ccessory. It is an accessory part of the vagus nerve.
XII. **H**ypoglossal. It innervates some muscles that move the tongue

Memory Jog

Oh, On **O**ctober **T**he **T**hirteenth, **A**unt **F**lorence **V**ester **G**ave **V**asquez **A H**ug

Use this sentence to remember the functions of cranial nerves as either **s**ensory, **m**otor, or **b**oth (somatic and autonomic motor fibers)

Sometimes **S**inging **M**elody **M**ay **B**ring **M**ixed (emotions) **B**ut **S**inging **B**lues **B**rings **M**ental **M**eltdown

Review Questions

1. The vital centers for control of the heart, respiration, and blood pressure are located in the _____.
2. The vomiting center is located in the _____.
3. The most important cardiac and vasomotor centers are situated in the _____.
4. The major control systems for balance and posture in the body as well as muscular coordination are located in the _____.
5. The area in the pons controlling the pattern of breathing is called the _____.
6. The midbrain structure(s) integrating auditory and visual reflexes is(are) the _____.
7. The red nucleus of the midbrain is the center for _____.
8. The limbic system of the brain is a collection of structures that are particularly important in _____.
9. List the parts of the limbic system. (1) Fornix (2) _____ (3) _____ (4) _____

10. The structure that contains neurons that transport hormones from the hypothalamus to the pituitary gland is the _____.
11. The center that serves as a relay station for olfactory neurons is the _____.
12. The intermediate mass, or massa intermedia, interconnects the _____.
13. The structure containing nuclei that regulate body temperature, water balance, appetite, and gastrointestinal activity is the _____.
14. The paraventricular nucleus and the supraoptic nucleus both seem to be involved in _____.
15. The part of the brain that links the nervous and endocrine systems is the _____.
16. The primary sensory area is located in the _____.
17. The auditory area is located in the _____.
18. The primary motor area is located in the _____.
19. The visual area is located in the _____.
20. Consciousness of heat, cold, and pressure are localized in the _____ of the cerebrum.
21. Disorders of memory and understanding are most commonly associated with lesions in the _____.
22. The groove that separates each temporal lobe from the lower portions of the frontal and parietal lobe is known as the _____.
23. The region of the cortex at the posterior part of the frontal lobe is called the _____.
24. Personality, morals, ethics, and general intelligence are located in the _____.
25. The corpus callosum interconnects the _____.
26. The raised area of the cerebrum is termed _____.
27. The connections between gyri of a single hemisphere are called _____.
28. Voluntary movements are initiated in the _____.
29. Hemispheres of the cerebrum are connected by nerve fibers called the _____.
30. The groove that separates the two cerebral hemispheres is called the _____
31. The cranial nerve that innervates the superior oblique muscle of the eye is _____.
32. The cranial nerve that controls the muscle of mastication is the _____.
33. The cranial nerve that innervates the trapezius muscle is _____.
34. The cranial nerve involved in hearing is _____.
35. Changes in the function of the cardiovascular system are caused by _____.

Chapter 14

Sensory-Motor Integration

14.1 Sensation or **perception** is the ability of the body to consciously detect and interpret the effects of stimuli on sensory receptors. The levels of sensation in the CNS are as follows:

1. Spinal cord: No immediate brain response is involved. Most reflexes are this level.
2. Lower brain stem: There is subconscious motor response.
3. Thalamus: There is a rough localization of stimulus origin, but limited identification of it.
4. Cerebral cortex: There is actual perception of the stimulus—specific localization/identification.

14.1.1 Types of sensations

Temperature, pain, pressure, balance, position, and **special senses**.

14.1.2 Senses

The brain uses senses to perceive information about the environment and the body. There are five senses: smell, taste, sight, hearing, and touch. These senses are now divided into two basic groups: general senses and special senses.

14.2 Classification of Receptors

14.2.1 Simple receptors are general senses for touch, pain, temperature, and proprioception. They are free nerve endings with no accessory structures. They are widely distributed. Examples of more complex free nerve endings are:

1. **Merkel (tactile) disks**: They are widely distributed at the basal layer of the epidermis just above the basement membrane. They detect the sensations of light touch and superficial pressure.

2. **Hair follicle receptors** (hair end organs): These are highly sensitive nerve endings that respond to slight bending of the hair and light touch. The sensation they elicit is not very well localized.

3. **Pacinian (lamellated corpuscles)**: These receptors are located deep in the dermis or hypodermis. They are responsible for sensing deep cutaneous pressure and vibration.

4. **Meissner's corpuscle**: These are highly discriminative receptors located in the dermal papillae. They are involved in two-point discrimination. They are very numerous in the tongue and fingertips.

5. **End organs of Ruffini**: These receptors are located in the dermis of the skin, mainly in the fingers. They respond to continuous touch or pressure.

6. **Muscle spindles**: These receptors are located in skeletal muscles, where they provide information about the length of the muscle. They play an important role in maintaining muscle tone of postural muscles.

7. **Golgi tendon organs**: They are located at junctions between tendon and muscle, where they monitor muscle contraction force.

14.2.2 Complex receptors detect sight, sound, taste, touch, temperature, vibration, and pain. Locations of receptors

- **Exteroceptors** are superficially located. They are sensitive to stimuli arising outside the body such as sound, smell, taste, touch, pain, pressure, temperature, and sight.

- **Interoceptors** are internally located receptors (visceroceptors). They monitor internal events, such as blood pressure, blood gases, and cardiac output. The sensations are usually not at a conscious level.

- **Proprioceptors** are located in muscles, tendons, joints, and the inner ear. They monitor body position and movements.

Classification according to the stimulus type that excites them:

1. **Mechanoreceptors**: physical distortion
2. **Nociceptors:** pain
3. **Photoceptors (photoreceptors)**: light
4. **Chemoceptors (chemoreceptors)**: chemical concentration

14.2.1 Mechanorecptors

These are cutaneous receptors that include **tactilereceptors**, **thermoceptors** and **nociceptors**.

14.3 Localization of Pain

- Somatic pain can be felt superficially in the skin or deep in joints, muscles or tendons.
- Phantom pain sensation is "felt" in absent limbs. It is caused by the trauma of amputation as a result of the stimulation of proximal parts of sensory nerves.
- Visceral pain sensation is felt in internal structures and can be localized or diffused.
- Referred pain is visceral pain felt distant to the affected organ.

14.4 Proprioceptors

These receptors allow awareness of body position and movement and state of muscle contraction. There are three major types of proprioceptors—muscle spindle, tendon organs, and joint kinesthetic receptors—and hair cells of the internal ear.

14.5 Organization of Sensory Pathways

Somatic Sensory Pathways

The somatic receptor sends inputs that cross over at the spinal cord or brain stem to the thalamus and then to the somatosensory cortex where conscious sensation is felt. There are two major ascending systems of tracts involved in conscious perception of external stimuli. They are the anterolateral system and the posterior column system.

Anterolateral System (pathway). The anterolatetral system includes:

- Spinothalamic tracts: carry sensations of pain, temperature, tickle, itch, and crude touch
- Spinocorticular tracts
- Spinomesencephalic tracts

Three neurons are involved in transmitting impulses along the spinothalamic tracts from the peripheral receptors to the cerebral cortex. They are:

1. Primary (first order) neuron: The primary neuron cell bodies are located in the dorsal root ganglia of the spinal cord. They relay sensory input from the periphery to the posterior horn of the spinal cord, where they synapse with (associative) neurons.
2. Secondary (second order) neuron: The secondary neurons cross to the contralateral side of the spinal cord through the anterior portions of the gray and white commissures and enter the spinothalamic tract, where they ascend to the thalamus.
3. Tertiary (third order) neurons: They are located in the thalamus and conduct impulses to the somatosensory cortex of the cerebrum.

Dorsal-Column/Medial-Lemniscal System

The dorsal column carries the sensations of two-point discrimination, proprioception, stereognosis, weight discrimination, pressure, and vibration. It divides into two tracts at the spinal cord based upon the source of the stimulus. The two tracts are:

1. Fasciculus gracilis conveys sensations from nerve endings below the midthorax level. It terminates by synapsing with the second order neuron in the nucleus gracilis in the medulla oblongata.
2. Fasciculus cuneatus conveys sensations from nerve endings above the midthorax. It terminates by synapsing with second order neurons in the nucleus cuneatus in the medulla oblongata.

Spinocerebellar System

The spinocerebellar tracts convey sensation involving subconscious muscle and joint movements. Two cerebellar tracts extend through the spinal cord.

1. **Posterior spinocerebellar tract:** It originates in the thoracic and upper lumber regions and contains uncrossed nerve fibers. It transmits proprioceptive information, such as equilibrium, posture, and coordination. It is composed of two-neuron sets.
2. **Anterior sinocerebellar tract:** It carries information from the lower trunk and lower limbs and contains crossed and uncrossed nerve fibers. It allows cerebellar regulation of posture, balance, and skilled movements.

14.6 Motor Pathways

Motor tracts are descending pathways containing axons that carry action potentials from the regions of the cerebrum or cerebellum to the brain stem or spinal cord. It is divided into two groups: direct pathways and indirect pathways.

14.6.1 Direct (Pyramidal) Pathways

Direct pathways include two tracts:

3. The corticospinal tract is involved in direct cortical control of movements below the head (muscles of the neck and trunk).
4. The **corticobulbar** tract controls voluntary movements of the head and neck.

14.6.2 Indirect Pathways

The major tracts are the rubrospinal, vestibulospinal, and reticulospinal tracts.

1. **Rubrospinal tract:** Its neurons begin in the red nucleus, decussate in the midbrain, and descend the lateral column of the spinal cord. It plays a major role in regulating fine motor control of muscle in the distal part of the upper limbs.
2. The **vestibulospinal tracts** are involved in the maintenance of upright posture. They originate in the vestibular nuclei and synapse with interneurons in the spinal cord.
3. The **reticulospinal tract** is found in the reticular formation of the pons and medulla oblongata. It maintains posture through the action of trunk and proximal upper and lower limb muscles.

Basal nuclei are also a major portion of indirect pathways. They interact with other indirect pathways to modify and refine motor activities.

14.7 Integrative Functions

14.7.1 Learning

- Ability to gain skills and knowledge
- Involves change in behavior
- Associated with rewards and punishment

14.7.2 Memory is divided into three major types: sensory, short-term, and long-term.

1. Sensory memory: Sensory memory lasts less than a second, and it involves transient changes in membrane potentials.
2. Short-term memory: If the given information is considered vital enough, it is moved into short-term memory. It may last seconds to a few minutes, depending on the number of bits received.
3. Long-term memory lasts for hours or a lifetime. It involves anatomical and biochemical changes in the brain. The hippocampus is involved in retrieving the actual memory, such as remembering a person's telephone number. The amygdala is involved in the emotional aspect of the memory.

14.8 Sleep and Wakefulness

Sleep is defined as a state of changed consciousness or partial consciousness, from which one can be aroused by stimulation. Electroencephalograms (EEG) display wavelike patterns called brain waves. They are classified as alpha, beta, theta, or delta waves.

1. Alpha waves are observed in a normal person who is awake but in a quiet, resting state with the eyes closed.
2. Beta waves are observed during intense activity.
3. Theta waves are usually seen in children and adults with certain brain disorders.
4. Delta waves occur in deep sleep and in patients with severe brain disorders.

Types of Sleep

The two major types of sleep are non-rapid-eye-movement (NREM) sleep and rapid-eye-movement (REM) sleep. They alternate through most of the sleep cycle.

Review Questions

1. The type of receptors that have large receptive fields is called n_____.
2. Endorphins can inhibit impulses initiated by _____.
3. Receptors that are scattered immediately beneath the surface of the skin is called t_____.
4. A tactile receptor that responds to deep pressure is a(n) _____.
5. Sensory receptors that monitor the position of joints are called _____.
6. Sensory receptors that respond to changes in blood pressure are called _____.
7. The spinal tract that carries sensory information concerning fine touch and pressure is the _____.
8. The spinal tract that relays information from proprioceptors to the CNS is the _____.
9. The spinal tract that relays information concerning pain and temperature to the CNS is the _____.
10. The spinal tract that relays information concerning crude touch and pressure to the CNS is the _____.
11. Interneurons of sensory pathways that are located in the spinal cord or brain are referred to as _____ neurons.
12. Sensory neurons that are located in the thalamus and project to the sensory cortex of the cerebrum are _____ neurons.
13. Explain why it is possible for an individual to distinguish sensations that originate in

different areas of the body.

_____

14. The spinal tract that regulates voluntary motor control of skeletal muscles on the same side of the body is the _____ tract.

15. The spinal tract that regulates voluntary motor control of skeletal muscles on the opposite side of the body is the _____.

16. The spinal tract that regulates involuntary control of posture and muscle tone is the _____ tract.

17. The spinal tract that controls involuntary regulation of reflex and autonomic function is the _____.

18. The spinal tract that controls involuntary regulation of eye, neck, and arm position in response to visual and auditory stimuli is the _____ tract.

19. What is the function of the pyramidal tract? _____

_____

20. An autonomic motor neuron whose cell body lies in the CNS is called a(n) _____ neuron.

21. Define receptor specificity.

_____

22. Define the term generator potential.

_____

_____

23. _____ are receptors in the lung that monitor the degree of lung expansion.

24. What would be the result of damage to the tectospinal tracts?

_____

25. The superior or inferior colliculi are located in the _____ or roof of the mesencephalon.

26. The _____ is a map of the sensory cortex of the cerebrum.

27. The structure concerned with maintaining the state of wakefulness is the _____.

28. Long-term memory depends on the formation of _____.

29. The ability to recognize by "feel" the size, shape, and texture of an object is called _____.

30. Acute pain is carried by A fibers, while chronic pain is carried by _____ fibers.

Chapter 15
The Special Senses

The special senses allow us to detect specific changes in our environment through a richer experience via smell (olfaction), taste (gustation), vision, hearing, and equilibrium.

15.1 Olfactory Sensation (Smell)

Olfactory receptors are classified as chemoreceptors. The impulse generated by these receptors projects to higher cortical areas and the limbic system.

- The receptors are innervated by facial nerve (VII)
- They are located in the upper nasal cavity made up of two layers:
  1). Olfactory epithelium that contains olfactory receptors with bipolar neurons that have a low threshold and that adapt rapidly. The epithelium contains two types of cells: (1) supporting cells and (2) basal cells, which are stem cells that replace old receptors.
  2). Olfactory (Bowman's Glands) that produce mucus that dissolves gases.
- Substances to be smelled must be volatile, water-soluble, and lipid-soluble.

15.1.1 Neural pathway

1. Receptor axons
2. Olfactory (I) nerve
3. Olfactory bulbs
4. Olfactory tract
5. Prepyriform cortex (Conscious perception of smell)
6. Limbic system: The medial olfactory area is involved in visceral and emotional response to odors.
7. Hypothalamus, amygdala, and other regions of the limbic system where emotional response to odors occur

15.2 Gustatory Sensation (Taste)

- The receptors for taste gustation are located in taste buds.
- Substances to be tasted must be in solution in saliva.
- Receptor potentials developed in gustatory hairs cause the release of neurotransmitter that give rise to nerve impulses.

15.2.1 Receptors

- Superior of tongue with epithelial projections called papillae containing taste buds

15.2.2 Cell types

1. Supporting (sustentacular)
2. Receptor cells—adapt rapidly
3. Basal cells—replace old receptors

15.2.3 Innervation

1. Anterior one-thirds of the tongue is innervated by facial nerve (VII).
2. Posterior one-third of the tongue is innervated by glossopharyngeal nerve (IX).
3. Throat and epiglottis are innervated by vagus nerve (X).

15.2.4 Neural Pathway

6.   Receptors
3.   Nerves VII, IX, X and their ganglion
4.   Medulla oblongata
5.   Limbic system
6.   Hypothalamus
7.   Thalamus
8.   Gustatory area of cerebral cortex

Note: Decussation takes place in the medulla oblongata and projects from there to the thalamus; projects to taste area of the cortex (extreme inferior end of the postcentral gyrus)

15.3 Vision

Visual receptors are contained in the eyes, where they detect light and create visual images.

15.3.1 Accessory Structures of the Eye

1.   Eyelids (palpebrae) provide shade, protect, and spread lubricants produced by Meibomian glands.
2.   Conjunctiva is a mucus membrane for protection.
3.   Eyelashes and eyebrows provide protection from foreign objects.
4.   Lacrimal (tear) apparatus secretes tears that clean and lubricate.
5.   Extrinsic eye muscles move the eye.
6.   Cilary glands (modified sweat gland)
7.   Meibomian glands (produce sebum)
8.   Lacrimal canaliculi collect excess excess tears through openings called punctum
9.   Lacrimal sac leads to nasolacrimal duct which opens into nasal cavity.

15.3.2 Structural Divisions of the Eyeball

Fibrous Tunic
- Outermost layer consisting the sclera (posterior) and cornea (anterior)
- The sclera and cornea contain an opening at their junction known as the scleral venous sinus.

Sclera
- It is referred to as the "white of the eye."
- It is made up of a white coat of dense fibrous tissue that covers the eyeball, except the most anterior portion, the iris.
- It gives shape to the eyeball and protects its inner parts.
- The posterior part is pierced by the optic (II) nerve.

Cornea
- It is a nonvascular, transparent, fibrous coat.
- It acts in refraction of light.

Vascular Tunic (Uvea)

Includes the iris, the ciliary body, and the choroid.

a.   **Iris**
- The iris is the colored portion seen through the cornea and consists of circular iris and radial iris smooth muscle fibers.
- The black hole in the center of the iris is the pupil, through which light enters the eyeball.
- The iris regulates the amount of light entering the posterior cavity of eyeball through the

117

action of two groups of smooth muscles. They are the circular **sphincter papillae** and the radial **dilator papillae.**

b. Choroid
  - The choroid absorbs light rays so they are not reflected and scattered within the eyeball.
  - It provides nutrients to the posterior surface of the retina.

c. Ciliary body
  - It consists of ciliary processes (epithelial lining cells that secrete aqueous humor that fills the anterior chamber) and ciliary muscle (which alters the shape of the lens).

Retina (Nervous Tunic)
  - It lines the posterior three-quarters of the eyeball and is the beginning of the visual pathway.
  - It functions in image formation.

Structure of the Retina

  - The pigmented layer aids the choroid in absorbing stray rays. They also act as phagocytes and store the vitamin A needed by photoreceptor cells.
  - The neural (nervous) layer contains millions of photoreceptor cells that transduce light energy. It contains three zones of neurons named in the order of nerve impulse conduction, namely, photoreceptor neurons, bipolar neurons, and ganglion neurons. The photoreceptor neurons are called rods or cones.
    1. **Rods** are specialized for black-and-white vision in dim light.
    2. **Cones** are specialized for color vision. Cones are densely concentrated at the central fovea in the center of the macula lutea.
  - The **macula lutea** is the exact center of the posterior portion of the retina, corresponding to the visual axis of the eye.
  - The **fovea centralis** is the area of sharpest vision because of the high concentration of cones. Rods are absent from the fovea and macula.
  - The optic disc (blind spot) lacks photoreceptors, and light focused on it cannot be seen.

Chambers of the Eye:
  - Anterior chamber is filled with aqueous humor produced by the ciliary processes.
  - Posterior chamber:is filled with vitreous humor which helps to maintain intraocular pressure, holds lens and retina in place and refracts light.

Photoreceptors consist of two parts:
  1. The outer segment is responsible for light transduction to an electrical signal.
  2. Inner segment: Action potentials are generated here.

Photopigments are constantly renewed, and they are responsible for initiating receptor potential.
Retinal combines with opsins to form visual pigments:
  1. Cis: Form in darkness
  2. Trans: Form in light

Retinal Processing of Visual Input
  - Receptor potentials develop in rods and cones.
  - Potentials spread to synaptic terminals.
  - Neurotransmitters are released.
  - Ganglion depolarizes.
  - Nerve impulse forms.

Visual Neural Pathway

- Light from each half of the visual field projects to the opposite side of the retina
- Optic nerve (II) (consisting of axons extending from the retina to the optic chiasm)
- Optic chiasm (Medial half fibers cross over, lateral half do not cross.)
- Optic tract consists of axons that have passed through the optic chiasm (with or without crossing) to the thalamus.
- The axons synapse in the lateral geniculate nuclei of the thalamus
- Optic radiations (formed by the neuron fibers of the thalamus that project to the visual cortex
- The right part of each visual field projects to the left side of the brain, and the left part of each visual field projects to the right side of the brain.)
- Visual areas of cerebral cortex (occipital lobes): Conscious perception of visual images occurs here.

  Binocular vision: The region of overlap consisting of areas seen with both eyes at the same time

  Accomondation: It is the process of changing the shape of the lens.

15.4 Anatomy of the Ear

The ear is divided into three anatomical regions: the external ear, the middle ear, and the inner ear.

15.4.1 The **external ear** collects and funnels sound.

Parts
1. The auricle (pinna) contains the helix and lobule.
2. External auditory canal
3. The tympanic membrane (eardrum) transfers vibrations.
4. Ceruminous glands produce cerumen.

15.4.2 Middle ear (tympanic cavity):

- Located in the cavity in temporal bone, filled with air
- Auditory (Eustachian) tube opens to the pharynx; it equalizes air pressure on the tympanium.
- Auditory ossicles
    1. The malleus (hammer) is attached to the tympanium.
    2. The incus (anvil) links the malleus to the stapes.
    3. The stapes (stirrup) vibrates against the oval window.

15.4.3 The **inner ear (labyrinth)** contains receptors that provide the senses of equilibrium and hearing. It contains a series of cavities called the bony labyrinth in the petrous portion of the temporal bone. It is divided into three areas:
1. Semicircular canals contain receptors for equilibrium.
2. The vestibule contain receptors for equilibrium.
3. The cochlea contains receptors for hearing.

15.3   Physiology of Hearing
1. Auricle collects sound waves.
2. The external auditory canal carries sound waves to the tympanium, which vibrates and sets ossicles in motion.
3. The stapes vibrates against the oval window, which vibrates, causing waves in the perilymph.

4. The vibrating stapes causes the fluid inside the cochlea to begin to move.
5. The endolymph moves the basilar membrane.
6. Spiral organ hair cells strike the tectorial membrane.
7. Receptor potential develops.
8. The nerve impulse travels to the brain via the cochlear branch (vestibulocochlear nerve (VIII))
9. A signal is sent to the brain for the interpretation of hearing.

15.5.1 Equilibrium
- Equilibrium is provided by the receptors of the vestibular apparatus.
- Semicircular canals convey information about rotational movements of the head.
- The saccule and utricle (otholitic organs) convey information about your position with respect to gravity.

15.5.2 Otolithic Organs
- Located at right angles on the walls of utricle and saccule
- Contain maculae receptors for static equilibrium—head position
- Help with dynamic equilibrium—changes in linear velocity
- Contain stereocilia and kinoclium hair cells

15.5.3 Semicircular Canals
The semicircular canals contain three ducts that lie at right angles to each other. Each duct has an enlarged swelling at one end called an ampulla, which houses receptor hair cells. The semicircular canals detect changes in rotational velocity.

15.5.2 The Utricle and Saccule
They provide equilibrium sensations whether the body is moving or not.

Review Questions

1. **What is the function of the basal gland of the olfactory epithelium?**

   _____

   _____

   _____

2. **Olfactory receptors normally react to what type of stimulus?**

   _____

3. **The first synapse of the olfactory neural pathway occurs in the _____.**
4. **The muscle that opens the eye is the _____.**
5. **The image of objects being closely examined fall on the _____.**
6. **The "white of the eye" is called the _____.**
7. **The fibrous tunic consists of the _____ and _____.**
8. **Cone cells are primarily found in the _____.**
9. **The structures that produce aqueous humor are known as the _____.**
10. **Which part of the retina lacks both rod and cone cells?**

    _____

11. **The portion of the eye concerned with image formation is the _____.**
12. **What causes the lens of the eye to thicken?**

    _____

13. **What causes excitation of the rods of the retina?**

    _____

14. **Rhodopsin is broken down by light into _____.**

120

15. When retinene and scotopsin combine together, _____ is formed.
16. _____ is formed directly from vitamin A.
17. The visual association areas are found within _____ lobe.
18. What is the function of the lateral geniculate?
   _____

19. Earwax is produced by the _____.
20. What is the function of the auditory (Eustachian) tube?

   _____

21. What is the function of the crista?
   _____

22. What is the function of the middle ear?
   _____

23. What is the function of the basilar membrane?
   _____

24. The pitch of sound is analyzed in the _____.
25. The vestibular apparatus is made up of the _____, _____, and

   _____.
26. What is hypermetropic?
   _____

27. The fluid within the membranous labyrinth is called _____.
28. The membrane that supports the organ of Corti is called the _____.
29. The clouding of the lens is called a(n) _____.
30. Conjunctivitis caused by Chlamydia is termed _____.

Chapter 16

The Autonomic Nervous System

1. The autonomic nervous system (ANS) regulates the activity of the smooth muscle, cardiac muscle, and certain glands.

2. It helps maintain homeostasis by receiving a continual flow of sensory afferent input from receptors in organs and sending efferent motor output to the effector organs.

3. The ANS is regulated by centers in the brain, mainly by the hypothalamus and medulla oblongata, which receives input from the limbic system and other regions of the cerebrum.

16.1 Anatomy of Autonomic Motor Pathways

1. Preganglionic neurons are myelinated; post-ganglionic neurons are unmyelinated.
2. Preganglionic neuron is the first of two motor pathways
   - Its cell body is in the CNS.
   - The preganglionic fiber passes out of the CNS as part of the cranial or spinal nerve, later splits from the nerve and extends to an autonomic ganglion, where it synapses with the postganglionic neuron.
3. The second neuron in the autonomic pathway is the postganglionic neuron, and it lies entirely outside the CNS.
   - Its cell body and dendrites are located in an autonomic ganglion, where it synapses with one or more preganglionic fibers.
   - The axon of a postganglionic neuron, the postganglionic fiber, is unmyelinated and terminates in a visceral effector.
4. The cell bodies of sympathetic preganglionic neurons are in the lateral gray horns of the 12 thoracic and first two or three lumbar segments.
5. The cell bodies of parasympathetic preganglionic neurons are in cranial nerve nuclei (III, VII, IX, and X) in the brain stem and lateral gray horns of the second through fourth sacral segments of the cord.
6. Autonomic ganglia are classified as sympathetic trunk (vertebral chain) and are lateral to the vertebral column, the prevertebral (collateral chain) ganglia are anterior to the vertebral column, and terminal (intramural) ganglia are close to or inside the visceral wall.
7. Networks of autonomic fibers are referred to as autonomic plexuses.

16.1.1 Physiology of Autonomic Nervous System

- Areas innervated by sympathetic nerves only: kidneys, sweat glands, adipose cells, arrector pili muscles, and most blood vessels
- The lacrimal gland is innervated by the parasympathetic system only.
- Other areas of the body are mostly innervated by both (dual innervations).

16.2 Sympathetic Division

1. Copes with stress and emergencies
2. Relates to energy using processes
3. Called to action by fear, rage, physical exertion
4. Fight or flight response—e.g., dilate pupils

5. Increase BP, CO, blood sugar, respiration, circulation to critical organs and muscles
6. Decrease circulation to skin and non-essential organs and digestion

16.3 Parasympathetic Division

Actions

Conserve, restore energy
1. Occur when body is resting or recovering
2. Controls: salivation, lacrimation, urination, and defecation
3. Decreases heart rate (also known as vagal massage)

16.4 Neurotransmitters and Receptors
16.4.1 Acetylcholine
- Acetylcholine is released by cholinergic fibers of ANS preganglionic neurons.
- Parasympathetic postganglionic neurons
- Some sympathetic postganglionic neurons innervate sweat glands, blood vessels, and skeletal muscle

Note: Cholinergic effects are brief. Acetylcholinesterase rapidly digests the acetylcholine.

16.5 Types of cholinergic receptors
16.5.1 Nicotinic
- Excitation of postsynaptic cells
16.5.2 Muscarinic
- Inhibits cardiac muscle
- Inhibits GI tract sphincters
- Excites iris of eye

16.5.3 Epinephrine/NE
- Adrenergic fibers release NE or epinephrine.
- Excites sympathetic postganglionic neuron

Note: Adrenergic effects last longer and cause greater divergence, slow uptake and slow breakdown by enzymes monoamine oxidase (MAO) or catechol-O-methyltransferase (COMT) in surrounding tissues.

Adrenergic receptors:
1. Alpha1 constricts blood vessels.
2. Alpha2 promotes blood clotting.
3. Beta 1 increases renin secretion, cardiac output, and lipolysis.
4. Beta 2 dilates deep blood vessels, constricts surface blood vessels, and relaxes smooth muscle.

16.6 Hypothalamic Control of the Nervous System
The main integration center of the nervous system is the hypothalamus. Posterior and lateral hypothalamic regions direct sympathetic functions, whereas anterior and medial areas direct parasympathetic functions.

Input goes to the hypothalamus regarding the emotional state, temperature, osmolarity, glucose, ions and gases blood levels, and special senses (taste and smell).

Note: Parasympathetic division is involved in SLUD response—S=Salivation, L=Lacrimation, U=Urination, D=Defecation.

Review Questions

1.   The second-order neurons of the autonomic nervous system are located in

     _____ _____.

2.   Postganglionic axons of autonomic neurons are usually myelinated/unmyelinated. (Choose one)

3.   What is the function of splanchnic nerves?

     _____

4.   The celiac ganglion innervates the _____.

5.   Sympathetic innervations of the urinary bladder and sex organs is by way of the

     _____.

6.   Sweat glands contain _____ receptors.

7.   Nicotinic receptors

     _____.

8.   Muscarinic receptors are activated by _____.

9.   What is dual innervation?

     _____.

10. Define declarative memories.

     _____.

11. What is meant by the term "memory consolidation"?

     _____.

12. What does dementia mean?

     _____.

13. Another name for the thoracolumbar division is the _____ division.

14. In the sympathetic division, preganglionic fibers are _____ and postganglionic fibers are _____.

15. Craniosacral division is another name for the _____ division.

16. _____ are the simplest functional units of the autonomic nervous system.

17. Drugs that mimic the activity of one of the normal autonomic neurotransmitters are called _____.

18. During _____ sleep, dreaming occurs.

19. Synapses that use acetylcholine as a transmitter are called _____.

20. Neurons that use norepinephrine as a transmitter are called _____.

Chapter 17

The Endocrine System

The endocrine system is responsible for long-term, body-wide coordination and development of cellular function. It is composed of glands that secrete chemical signals (hormones) into the circulatory system.

17.1 Actions of the Endocrine System

Hormones are released in low concentration into extracellular space.

They diffuse into the blood and travel in the blood in two ways:

> (1) Water-soluble hormones travel in free form—e.g., peptides, proteins, and catecholamine
>
> (2) Lipid-soluble hormones travel bound to proteins—e.g., steroids and thyroid hormones

17.1a Hormone Interaction with Target Cell Receptors

Hormones bind with two types of receptors:

17.1.1 Plasma membrane receptors (membrane-bound receptors)

These are used by large molecules and water-soluble molecules that cannot pass through the plasma membrane. Examples of such molecules are catecholamine, peptide, and protein hormones. The receptor-hormone complex formed activates a second messenger inside the cell and produces a receptor cascade.

Intracellular receptors

These are used by lipid-soluble hormones, such steroids and thyroid hormones. They diffuse through the plasma membrane and bind to the receptors in the cytoplasm or in the nucleus of the cell. The receptor-hormone complex formed interacts with DNA transcription and alters gene expression. The presence or absence of a hormone can also affect the number and nature of hormone receptor proteins in the cell membrane.

Up regulation: The target cells form more receptors in response to rising levels of a specific hormone.

Down regulation: The target cells form fewer receptors as a result of prolonged exposure to high concentration of a hormone.

17.2 Patterns of Hormonal Interactions

When a cell receives instructions from two hormones at the same time, four outcomes are possible.

They are as follows:

1. **Antagonistic:** Two hormones have opposing actions—e.g., PTH and calcitonin, or insulin and glucagon.
2. **Synergistic:** Two or more hormones are required for one effect—e.g., the glucose-sparing action of GH and glucocorticoid.
3. **Permissive:** Exposure to one hormone enhances the action of the second hormone. For example, epinephrine does not change energy consumption unless thyroid hormone is also present in a normal concentration.
4. **Integrative:** Two hormones may produce different, but complementary results in specific tissues and organ—e.g., the different effects of calcitrol and parathyroid

hormone on tissues involved in calcium metabolism.

17.3 Chemical Classes of Hormones

Steroids hormones are derived from cholesterol. Examples are aldosterone, cortisol, androgens (testosterone), calcitro, estrogen, and progesterone.

Peptides and proteins are derived from amino acids. Examples are releasing hormones, glucagon, oxytocin, ADH, insulin, calcitonin, and parathyroid hormones.

Eicosanoids are derived from arachidonic acid. Examples are prostaglandins and leukotrienes.

17.4 Mechanisms of Hormone Actions

17.4.1 Hormonal Action by Plasma Membrane Receptors
- This method is used by amino acid-based hormones that cannot penetrate the plasma membrane of target cells.
- The hormone (first messenger) binds to its receptor.
- The receptor changes shape and binds to G proteins.
- The G protein is activated as it binds GTP, which displaces GDP.
- Many G proteins within the cell membrane are activated.
- Adenylate cyclase molecules are produced and then generate the second messenger cAMP from ATP.
- cAMP levels increase greatly.
- Each cAMP activates many molecules of protein kinases (enzymes that transfer phosphate groups).
- Each protein kinase changes the function of several protein molecules.
- Cell functions are altered; the initial signal is greatly magnified, leading to the appearance of thousands of second messengers in the cell (i.e., amplification).
- The increase is short-lived; phospodiesterase (PDE) inactivates cAMP by converting it to adenosine monophosphate (AMP).

17.4.2 Hormonal Action by Intracellular Receptors
- This is used by lipid-soluble hormones, such steroid and thyroid hormones.
- They diffuse through plasma membrane.
- They bind to a carrier protein to get through the cytoplasm.
- They enter the nucleus and bind specific a receptor on the DNA.
- Gene expression is altered.
- New mRNA and protein are made.
- The new protein changes cell functions.

17.5 Control of Endocrine Activity

The endocrine glands of the body are stimulated to manufacture and release their hormones by three major types of stimuli: humoral, neural, and hormonal.

17.5.1 Humoral Stimuli

Some endocrine glands secrete their hormones in direct response to a change in blood levels of certain ions and nutrients. There is no direct nerve control. The response to stimuli can be a positive feedback or a negative feedback.

Positive feedback

- Stimulus is detected.
- Hormone is released.
- Hormone reaches target tissue.
- Hormone binds to receptor.
- Hormone induces effect.
- Feedback to endocrine structure occurs.
- More hormone is released.

Example: Elevated level of oxytocin triggers uterine contraction. The greater the force of contraction, the more oxytocin that is released.

Negative feedback:

- Stimulus is detected.
- Hormones are released.
- Hormones reach the target tissue.
- Hormone binds to receptor.
- Hormone produces desired effect.
- Hormone release is shut down.

Example: Parathyroid glands directly monitor calcium levels; when they detect decline from normal Ca^{2+} values, they secrete PTH, which reverses the decline and PTH secretion stops.

17.5.2 Neural Stimuli

Nerve fibers stimulate hormone release, which causes inhibiting and releasing factors to be released. **Examples:** The sympathetic nervous system stimulates the adrenal gland to release catecholamines (epinephrine and norepinephrine) during periods of stress. Also, oxytocin and antidiuretic hormones are released from the posterior pituitary gland in response to nerve impulses from hypothalamic neurons.

17.5.3 Hormonal Stimuli

Many endocrine glands release their hormones in response to hormones from other glands. For example, anterior pituitary gland hormones are regulated by releasing and inhibiting hormones produced by the hypothalamus. Many anterior pituitary hormones in turn stimulate other endocrine glands to release their hormones into the blood.

17.6 Release of Hormones from Posterior Pituitary Gland

- Neurohypophysis releases but does not synthesize oxytocin (OT) and antidiuretic hormone (ADH).
- OT and ADH are synthesized by neurosecretory cells in hypothalamic nuclei (paraventriclar and supraventricular).
- Hormones are packed into secretory vesicles.
- The vesicles move to neurohypophysis through the supropticohypophyseal tract by fast axonal transport. The vesicles stop at axon terminals in neurohypophysis.
- The nerve impulse travels down the supraopticohypophyseal tract.
- Hormones are released from vesicles by exocytosis.
- The hormone diffuses into the blood through the hypophyseal portal system.

17.6.1 Neurohypophysis Hormones

1. **Oxytocin OT (Pitocin)** stimulates smooth muscle contractions in the uterus and mammary glands. It is regulated by uterine stretching and suckling in consonance with estrogen. It is inhibited by progesterone.
2. **Antidiuretic Hormone ADH (Vasopressin)** acts on kidney tubules by decreasing urine output and increasing water retention and blood pressure. It acts by constricting arterioles. ADH release is regulated by stress, blood volume, drugs, osmoreceptors, and circadian rhythm.

Note: Diabetes insipidus is as a result of hyposecretion of ADH. Symptoms include excessive urination, dehydration, and thirst.

The **pituitary gland (hypophysis)** is suspended from the base of the brain by infundibulum and sheltered by the sella turcica of sphenoid bone. It is divided into posterior and anterior lobes on the basis of function and developmental anatomy.

The **anterior pituitary gland** is referred to as the adenohypophysis. It is linked to the hypothalamus by the portal system (superior hypophyseal artery, primary plexus, hypophyseal portal veins, and secondary plexus). The anterior pituitary gland is under the control of the hypothalamus.

17.6.2 Hormones of the Anterior Pituitary Gland

The hormones secreted by the anterior lobe of the pituitary gland are known as tropic hormones because they "turn on" endocrine glands or support the functions of other organs.

Growth hormone (hGH)

It is the most produced adenohypophysis hormone.

1. It is not a true tropic hormone because its effects are widespread and not limited to specific endocrine targets.
2. hGH increases the mitotic rate and cell growth.
3. It changes metabolism by increasing protein synthesis and stimulates breakdown of stored triglycerides by adipocytes, which stops many tissues from breaking down glucose, but starts breaking down fatty acids to generate ATP (glucose-sparing effect)
4. The production of GH is regulated by a growth hormone–releasing hormone, somatocrinin (GH-RH), and growth hormone–inhibiting hormone, somatostatin (GH-IH).

Note: Elevation of blood glucose levels by GH is referred to as the **diabetogenic effect**.

Associated disorders

1. Hyposecretion: pituitary dwarfism
2. Childhood hypersecretion: giantism
3. Adulthood hypersecretion: acromegaly

Prolactin (PRL) stimulates milk production in females. Hypersecretion leads to galactorrhea and amennorhea in females, impotence and infertility in males. Its secretion is regulated by prolactin-releasing and inhibiting hormones PRH and PIH

Thyroid-stimulating hormone stimulates the synthesis and secretion of thyroid hormones. It is regulated by TRH by monitoring blood levels of thyroid hormone, blood glucose, and basal metabolic rate. Stress and pregnancy also trigger TRH release. Somatostatin blocks all anterior pituitary hormones.

Follicle-stimulating hormone (FSH) is a gonadotropin that stimulates sperm development in testes in males and ova development in females. It also stimulates estrogen secretion. FSH is regulated by

GnRH from the hypothalamus.

Leuteinizing hormone (LH) is an interstitial cell–stimulating hormone (ICSH) that stimulates testosterone secretion in males, estrgen and progesterone secretion in females. Its regulation is under the control of GnRH.

Melanocyte-stimulating hormone (MSH) may influence post-natal skin pigmentation. It is regulated by melanocyte-releasing hormone (MRH)

Adrenocorticotropic hormone (ACTH) controls the synthesis and secretion of adrenal cortex hormones. Its secretion is regulated by hypothermia, pregnancy, stress, and corticotrophin-releasing hormone (CRH).

17.7 The Thyroid Gland

The hyroid gland contains two types of cells.

- **Follicular cells** release triiodothyronine (T3) and thyroxine (T4).
- **Parafollicular cells** release calcitonin (CT).

Steps in Synthesis and Secretion by Follicular Cells

1. Use iodide pump to trap iodide ions.
2. Make thyroglobulin glycoprotein chain.
3. Thyroglobulin exported to colloid
4. Pairing process performed by peroxidase, which converts iodide to iodine
5. TSH stimulates iodide transport into follicle cells and the production of thyroglobulin.
6. Follicle cells remove thyroglobulin from the follicles by pinocytosis. Thyroid hormones released
7. Hormones bind to thyroxine-binding globulin (TBGs) in blood.

Actions of T3/T4

Thyroid hormones enter target cells by means of an energy-dependent transport system, and they affect almost every cell in the body. They control metabolism and energy balance, growth and development, and oxygen uptake by tissue cells. Their release is stimulated by the following factors:

1. Low blood levels of thyroid hormones
2. Hypothermia
3. Hypoglycemia
4. Pregnancy
5. Altitude

Calcitonin helps to maintain homeostasis of calcium and phosphate ions by increasing kidney secretion of calcium ions. It increases osteoblast activity and decreases osteoclast activity. Its release is regulated by blood calcium levels.

17.8 The **Parathyroid Gland** is made up of two pairs of glands embedded in the posterior surface of the thyroid gland. It secretes parathormone (PTH), which maintains the homeostasis of some ions in the blood. It raises calcium and magnesium ions levels in the blood and lowers phosphate ion levels. It increases osteoclast activity and decreases osteoblast activity. Parathormone also activates vitamin D.

17.9 The **Adrenal (Suprarenal) Gland,** a highly vascular gland, is located on the superior border on each kidney.

The adrenal gland is divided into two parts with separate endocrine functions: a superficial adrenal cortex and an inner adrenal medulla.

Cortex

- Zona glomerulosa produces mineralocorticoids: aldosterone.
- Zona fasciculata produces glucocorticoids: hydroxycortisone, cortisone, and corticosterone.
- Zona reticularis produces gonadocorticoids: estrogens and androgens.

Medulla
- Produces epinephrine and norepinephrine

Mineralocorticoids

These are steroid hormones that affect the electrolyte composition of body fluids. Aldosterone is the principal mineralocorticoid produced by the adrenal cortex.

Aldosterone

1. **It acts** on kidney tubules.
2. It increases reabsorption of Na, Cl, and HCO3.
3. It increases excretion of K and H.
4. It reduces urine output.
5. It increases blood volume and blood pressure.
6. Its secretion is controlled via the renin-angiotensin pathway.

Glucocorticoids

They have effects on glucose metabolism. They cause an increase in glucose production from non-carbohydrate products **(gluconeogenesis).** When they are stimulated by ACTH from the anterior lobe of the pituitary gland, the zonal fasciculata secretes primarily cortisone.

Cortisol increases fatty acid use, protein breakdown, gluconeogenesis, and blood glucose. It also increases sensitivity to vasoconstriction and resistance to stress. It decreases inflammation, speeds up wound healing, and increases glucose uptake by cells. Its secretion is regulated by stress, glucocorticoid level, and CRH/ACTH.

Gonadocorticoid

The zona reticularis normally produces small quantities of androgens, the sex hormones produced in large amount in the testes in males, under stimulation of ACTH. The androgens are then converted to estrogens, the dominant sex hormones in females. These hormones assist in the prepuberal growth spurt and support the libido (female sex drive). They also promote increase in muscle mass and blood cell formation in females.

Epinephrine and Norepinephrine

Epinephrine makes up 75–80 percent of the secretions from the adrenal medullae, the rest being norepinephrine. They are involved in the "fight-or-flight" syndrome. They are sympathomemetic (linked to the sympathetic nervous system). They cause cells to increase their metabolic rates by mobilizing more glucose into the blood stream during periods of emergency. They increase blood pressure, cardiac output, and cellular metabolism. They dilate visceral blood vessels and the bronchial tree and constrict peripheral blood vessels and decrease digestion.

17.10 Ovaries and Testes

a. Ovaries produce sex hormones (estrogens and progesterone) related to the development and maintenance of female sexual characteristics, the reproductive cycle, pregnancy, lactation, and normal reproductive functions. The ovaries also produce inhibin and relaxin.

b. Testes produce sex hormones, testosterone, and inhibin responsible for the development of male secondary sexual characteristics and normal reproductive functions.

17.11 Pancreas

It performs exocrine and endocrine functions.

Exocrine: Acinar cells produce digestive enzymes.

Endocrine: Pancreatic islets produce several hormones as follows:

1. **Alpha cells** produce glucagon.

2. **Beta cells** secrete insulin.
3. **Delta cells** secrete somatostatin.
4. **F cells** secrete pancreatic polypeptide.

Glucagon acts on the liver when blood glucose is low. It stimulates glycogenolysis (breakdown of glycogen in the liver). It also stimulates gluconeogenesis (synthesis of glucose in the liver). It also stimulates beta oxidation of fats. Glucagon release is regulated by blood glucose level, diet, and somatostatin.

Insulin acts on the liver when blood glucose is high. It causes protein channels within the membrane to form pores, thereby letting glucose enter the cell for metabolism. Insulin release is regulated by blood glucose levels, hGH, ACTH, somatostatin, and gastrointestinal hormones.

Pancreatic polypeptide controls the release of pancreatic digestive enzymes.

17.12 The **Pineal Gland** is a part of the epithalamus. It lies in the posterior portion of the roof of the third ventricle. It secretes melatonin in a diurnal rhythm linked to the dark-light cycle.

17.13 The **Thymus Gland** produces thymopoietins and thymosins essential for the development of T lymphocytes (T cells) and immune response.

17.14 Other Hormone-producing Structures

1. Heart. The atria of the heart contain some specialized cardiac muscle cells that secrete atrial nitriuretic peptide (ANP), which reduces blood volume, blood pressure, and blood sodium concentration. It stimulates the kidney to increase the production of salty urine and inhibits aldosterone release by the adrenal cortex.

2. Gastrointestinal tract. The mucosa of the gastrointestinal tract contains hormone-secreting cells that release several amine and peptide hormones that help regulate a wide variety of digestive functions.

3. Placenta. The placenta secretes several steroid and protein hormones that play vital roles in pregnancy. Examples are estrogens, progesterone, and human chorionic gonadotropin (hGC).

4. Kidney. The kidney cells secrete erythropoietin, a protein hormone that signals the bone marrow to increase the production of red blood cells.

5. Skin. The skin produces cholecalciferol, an inactive form of vitamin D, when exposed to the sun. The compound then enters the blood, where it is modified in the liver to active vitamin D3.

Review Questions

1.   **What is the role of adenylate cyclase in hormone activity?**
_____

2.   **What are tropic hormones?**
_____

3.   **What is the role of cyclic AMP in endocrinology?**
_____

_____

_____

4.   **Describe the sequence of events that follow after a steroid hormone has reached a target cell.**

_____

_____

5.   **Describe the sequence of events that follow after thyroid stimulating hormone (TSH) has reached the target cell.**

6. The substance that acts as a "second messenger" inside of a cell is called
   _____.

7. The hormone that increases skin pigmentation is called _____.

8. Hypersecretion of growth hormone in an adult will result in _____.

9. Which hormones control the anterior lobe of the pituitary gland?
   _____

10. Which hormone stimulates the secretion of testosterone in the male?
    _____

11. Which hormone promotes protein synthesis?
    _____

12. How do hypothalamic-releasing hormones reach the pituitary gland?
    _____

13. Undersecretion of adrenocorticotropic hormone (ACTH) will lead to symptoms of
    _____ disease.

14. The neurohypophysis of the pituitary is functionally connected to the brain by the
    _____.

15. An increased amount of antidiuretic hormone (ADH) will lead to _____.

16. What causes diabetic insipidus?
    _____

17. What is the target tissue for oxytocin (OT)?
    _____

18. The target tissue for antidiuretic hormone (ADH) is the
    _____.

19. What is the cause of cretinism?
    _____

20. Name the cause(s) of myxedema.
    _____

21. The cause of Grave's disease is
    _____.

22. The antagonistic hormones that regulate blood calcium levels are _____
    and _____.

23. Mineralocorticoids are produced by the _____.

24. What is the primary stimulus for the secretion of aldosterone?
    _____

25. What causes Cushing's syndrome?
    _____

26. Delta cells of the pancreas produce _____.

27. Glycogenesis is controlled by _____.

28. What is the action of glucagon?
    _____

29. Glycogenolysis is stimulated by _____ and _____ hormones.

30. Why does alcohol consumption increase urine production?
    _____

    _____

    _____

Chapter 18
The Cardiovascular System: Blood

18.1 Physical Characteristics
1. Blood has viscosity greater than that of water.
2. The pH of blood is 7.35–7.45
3. Blood constitutes about 8 percent of body weight.
4. The average volume of blood ranges from four to six liters.

18.2 Functions of blood
The distribution functions of blood include:

- Delivering oxygen from the lungs and nutrients from the digestive tract to all body cells
- Transporting metabolic waste products from cells to elimination sites
- Transporting hormones to their target organs

The regulatory functions of the blood include:
- Maintaining appropriate temperatures by absorbing and distributing heat throughout the body
- Maintaining normal pH in body tissues
- Maintaining adequate fluid volume in the circulatory system

The protective functions of the blood include:
- Preventing blood loss
- Preventing infection

18.3 Plasma Composition
Blood plasma is a straw-colored, sticky fluid. It contains mostly water, with over 100
different dissolved solutes.
Composition:
- Proteins: **albumin, globulin, and fibrinogen**
- Nutrients: **amino acids, fatty acids, and glucose**
- Wastes **(nonprotein nitrogenous byproducts):** urea, uric acid, creatine, reatinine, and bilirubin
- Dissolved gases: **carbon dioxide, nitrogen, and oxygen**
- Electrolytes: Cations—**sodium, potassium, magnesium, and calcium**
          Anions—**chloride, sulfate, phosphate, bicarbonate**

18.4 Hemopoiesis or Hematopoiesis
Blood cell formation is referred to as hematopoiesis or hemopoiesis. This process occurs in the red
bone marrow and lymphoid tissue.
18.4.1 The **Hematopoietic stem cell or hemocytoblast** is located in the red bone marrow. These cells are
committed to a specific blood cell pathway by the action of specific hormones that stimulates membrane
surface receptors to specialize as follows:
Erythrocyte Production (erythropoiesis)
1. Myeloid stem cell
2. Early erythroblast

3. Late erythroblast
4. Normoblast
5. Reticulocyte
6. Erythrocytes

The entire process from hemocytoblast to erythrocyte formation takes three to four days.

Regulation of Erythropoiesis

- A drop in normal blood oxygen levels causes the kidney or liver to release erythropoietin. Erythropoietin (EPO) stimulates committed red bone marrow cells to rapidly mature into erythrocytes.
- Other required raw materials include proteins, iron, lipids, and carbohydrates.

Fate and Destruction of Erythrocytes

- Hemoglobin is broken down into its components (heme and globin chains).
- The globin chains are broken into amino acids and recycled.
- The heme is converted to biliverdin, which further converts to bilirubin.
- The iron portion of heme is released and recycled in the bone marrow.

Leukopoiesis: white blood cells
Thrombopoiesis: thrombocytes

18.5 Erythrocytes (Red Blood Cells)

1. Erythrocytes transport oxygen from the lungs to the various tissues of the body and transport carbon dioxide from the tissues to the lungs.
2. Mature cells have no nucleus.
3. They are flexible due to the cell membrane containing spectrin—a network of proteins.
4. Erythrocytes cells are capable of lining up in single file in the capillaries—referred to as a Rouleaux formation.
5. Their life span is about 120 days.
6. There are 4.3–5.2 million cells per cubic millimeter.
7. Erythrocytes make up about 35–45 percent of blood volume. This is referred to as hematocrit.
8. Erythrocytes contain hemoglobin—about 12–20 mg/ml (varies with age)
9. **Globin** has four protein chains/amino acids. Globin binds to carbon dioxide to form (carbaminohemoglobin).
10. **Heme** contains iron that binds to oxygen to form oxyhemoglobin.

18.6 Leukocytes (White Blood Cell)

1. They are nucleated without hemoglobin.
2. They have a variable life span.
3. There are 4,000–11,000 WBCs per cubic millimeter of blood.
4. They are able to slip out of capillaries and blood vessels (diapedesis).
5. They are able to follow a chemical trail of molecules released by damaged cells (chemotaxis)

Groups

Leukocytes are grouped into two major categories on the basis of structural and chemical characteristics: granulocytes and agranulocytes.

18.6.1 Granulocyte

Neutrophils are wandering and fixed macrophages that develop from monocytes.

Eosinophils combat the effects of histamine in allergic reactions, phagocytize antigen-antibody complexes, and combat parasitic worms.

Basophils develop into mast cells that liberate heparin, histamine, and serotonin in allergic reactions that intensify inflammatory responses.

18.6.2 Agranulaocytes

B lymphocytes differentiate into tissue plasma cells, which produce antibodies in the presence of antigens.

a.   T cells attack viruses and cancer cells.

b.   B cells give rise to plasma cells, which produce antibodies.

Monocytes differentiate into highly mobile macrophages.

Leukocytes Disorders

- **Leukopenia:** abnormally low white blood cell count (less than 5K /cubic mm)
- **Leukocytosis:** infection-related elevation of WBC count
- **Leukemia:** pathological elevation of WBC count—greater than 100K/cubic mm

Thrombocytes (platelets) are not cells but are cytoplasmic fragments. There are about 250–400 K fragments per cubic mm. They are essential for the blood-clotting process.

18.7 Hemostasis

Hemostasis is the stoppage of bleeding. The stages of hemostasis are as follows:

In a **vascular spasm,** narrow, sharp pressure on the smooth muscle of a blood vessel causes the wall to constrict to stop bleeding. It is most effective in the smaller blood vessels. The process may last up to 20 to 30 minutes, allowing time for platelet plug formation and blood clotting to form

Platelet plug formation involves the clumping of platelets around the damaged vessel to stop bleeding. The damaged blood vessel walls reveal sticky sites, and platelets adhere to the exposed collagen. Platelet granules begin to break down and release several chemicals like serotonin, ADP, thromboxane A2, and fibrin stabilizing factors, which help to form platelet aggregation.

Coagulation (clotting) involves the formation of a prothrombin activator, which converts prothrombin into thrombin. Thrombin then catalyzes the conversion of fibrinogen to a fibrin mesh.

Initiation of Coagulation

Clotting may be initiated by either the intrinsic or the extrinsic pathway, and both pathways are usually triggered by the same tissue-damaging events. The formation of prothrombinase is initiated by the interplay of two mechanisms called extrinsic and intrinsic pathways of blood clotting.

Extrinsic Pathway

- Occurs in seconds
- The vessel endothelium ruptures, exposing underlying tissues (e.g., collagen).
- Tissue factor (TF) or thromboplastin enters from outside the blood.
- Exposure of throboplastin to additional factors released by damaged tissues creates a "shortcut"

   by bypassing several steps of the intrinsic pathway.

Intrinsic Pathway

- Occurs in minutes
- All factors needed for clotting are present in the blood.
- Each pathway requires ionic calcium.
- The intermediate steps of each pathway cascade toward a common intermediate, factor X. Factor X combines with calcium ions, tissue factor of PF3, and factor V to form prothrombin activator. A clot forms in 10–15 seconds..

Common Pathway to Thrombin

Prothrombinase converts prothrombin to thrombin.

1.   Thrombin increases prothrombinase production (positive feedback).

2. Thrombin activates more platelets.
3. Soluble fibrinogen is converted to insoluble fibrin.
4. Fibrin locks up thrombin.
5. Thrombin also activates factor XIII (fibrin-stabilizing factor) in the presence of calcium ions.

Clot Retraction (Syneresis)

Tissue plasminogen activator (t-PA) activates plasminogen to plasmin. Plasmin dissolves (hydrolyzes) fibrin by the process called fibrinolysis and inactivates clotting factors.

Regulation of Clotting

- Structural and molecular characteristics of the endothelial lining of blood vessels—must be smooth and intact.
  - Antithrombic substances—heparin and protein C—secreted by endothelium prevent platelet adhesion.
  - Vitamin E is also regarded as a potent anticoagulant when it reacts with oxygen.

18.8 Human Blood Group

People have different blood types, and transfusion of incompatible blood can be fatal. Red blood cell membranes bear highly specific glycoproteins (antigens) at their external surfaces, which identify each of us as unique from others.

- Agglutinogen is a hereditary glycoprotein located on exterior of red blood cells.
- Agglutinin is an isoantibody that binds to agglutinogens not present in one's own body (agglutination reaction).

18.8.1 ABO Blood Groups are based on the presence or absence of two agglutinogens, type A and type B. Type O has neither agglutinogen, type AB has both antigens, and the presence of either A or the B agglutinogens result in group A or B respectively. The preformed antibodies in the plasma (agglutinins) act against red blood cells carrying ABO antigens that are not present on a person's own red blood cells.

18.8.2 Rh Blood Groups

- Rh positive factor is dominant to Rh negative.
- Rh negative is recessive.
- Anti-Rh develops only in an **Rh negative person after exposure to Rh+ blood**.
- Hemolytic disease of the newborn (HDN) occurs only in an Rh- female with an Rh+ fetus (erythroblastosis fetalis).
- HDN occurs only in a subsequent pregnancy.

18.9 Disorders: Homeostatic Imbalances

Anemia is a condition in which the oxygen-carrying capacity of the blood is reduced. The types of anemia are as follows:

Hemorrhagic anemia occurs as a result of major blood loss, such as GI bleeding, heavy menses, or major trauma.

Nutritional anemia results from the deficiency of iron, amino acids, and vitamin B12.

Pernicious anemia occurs when intrinsic factors are absent or vitamin B12 is not absorbed.

Hemolytic anemia is caused by broken red blood cell membranes, erythroblastosis fetalis, ingestion of toxic medication, or in individuals with thalasemia disease.

Aplastic anemia is caused by loss of bone marrow function, toxins, or radiation.

Sickle cell is a congenital disease caused by abnormally hemoglobin production.

Review Questions

1. The red blood cells that are released by the bone marrow into the blood stream are called _____.
2. Increase in white blood cells is called _____.
3. Which white blood cells produce defensins? _____.
4. The ability of white blood cells to reach an infected tissue depends on the process called _____.
5. The term that means lower white blood cells than normal is called _____.
6. Name the three granulocytes. (1) _____ (2) _____ (3) _____
7. The leukocyte that releases histamine is called a(n) _____.
8. The leukocyte that is converted to macrophages in the area of damaged tissue is called a(n) _____.
9. The engulfment of foreign organisms by certain types of blood cells is known as _____.
10. The cell type that aids in syneresis is called _____.
11. The substance that renders thrombocytes or platelets "sticky" is _____.
12. Platelets or thrombocytes are formed from stem cells called _____.
13. The conversion of soluble fibrinogen into insoluble fibrin requires an enzyme called _____.
14. Which factor converts prothrombin to thrombin? _____
15. Which substance enzymatically decomposes fibrin and reduces a clot? _____
16. A blood clot that forms in a blood vessel in the leg and finds its way to a blood vessel in the lung is described as a(n) _____.
17. Type AB blood contains _____ in the plasma.
18. Type O blood contains _____ on the cell surface.
19. A blood clot formed in an unbroken vessel is called a(n) _____.
20. The movement of phagocytes toward a particular chemical is called _____.

Chapter 19
The Cardiovascular System: The Heart

The heart is located in the ventral cavity, inside the thoracic cavity. It is enclosed in the pericardial cavity. It is enclosed within the mediastinum and two-thirds shifted to the left. The base of the heart is at the posterior surface and is predominantly in the left atrium. The apex containing mostly the left ventricle points toward the left.

19.1 Pericardial Sac

The heart is enclosed in a double-walled sac called the pericardium. The loosely fitting superficial part of the sac is the fibrous pericardium, which protects and anchors the heart. Deep to the fibrous pericardium is the serous pericardium, composed of two layers:

1. Parietal layer: lines the internal surface of the pericardium
2. Visceral layer (epicardium): integral part of the heart

The pericardial cavity is found between the two layers and contains serous fluid.

19.1.1 Cardiac Wall

- The epicardium is the visceral pericardium.
- The myocardium is the muscle layer.
- The endocardium lines the heart chambers.

19.1.2 Atria

- They have thin walls made up of pectinate muscle
- The walls are separated by the interatrial septum, which bears a shallow fossa ovalis (foramen).
- The right atrium receives oxygen-depleted blood from the body. This is the point of lowest BP in the body.
- The left atrium receives oxygen-enriched blood from the lungs.

19.1.3 Ventricles

- They have thick walls.
- Internal walls are marked by irregular ridges of muscle called trabeculae carnae.
- The chambers are separated by the interventricular septum.
- The right ventricle receives blood from right atrium and pumps it to the lungs.
- The left ventricle receives blood from the left atrium and pumps it to the rest of the body.
- The left ventricle has the highest blood pressure in the body.

19.1.3 The Heart Valves

Blood flows through the heart in one direction—from atria to the ventricles and out the great arteries—leaving the superior aspect of the heart. This one-way traffic is enforced by four heart valves. Attached to each valve flap are cords called chordae tendinae that anchor the valves to the papillary muscle that prevents the reverse flow of blood.

- Tricuspid valve: Right atrioventricular valve, located near the sternum
- The pulmonary semilunar is found in the right ventricle; it guards the base of the pulmonary trunk .
- Bicuspid (mitral): Left atrioventricular
- The aortic semilunar valve guards the base of the aorta issuing from the left ventricle.

19.3   Conduction System

1. Sinoatrial (SA) node
2. Atrioventricular (AV) node
3. Atrioventricular (AV) bundle (bundle of His)

4. Right and left bundle branches
5. Purkinje fibers

19.5 Physiology of Cardiac Muscle Contraction

Electrical cardiac fibers transmit impulses to contractile fibers.

Depolarization
- Rapid depolarization from resting potential of -90mV
- Voltage-gated fast Na channels open.
- Voltage-gated K channels begin to open.

Plateau
- Depolarization maintained 250 times longer than in skeletal muscles
- Cardiac contraction strength can be adjusted by changing Ca flow. Increased Ca inflow increases contraction strength; decreased Ca inflow decreases strength.

Repolarization
- Voltage-gated K channels open.
- Voltage-gated fast Na channels close.
- Voltage-gated slow calcium channels close.
- Resting membrane potential is reestablished.

Refractory period
- Second contraction cannot be initiated at refractory period.
- It lasts longer than contraction.

19.6 Electrocardiography

The electrical currents generated and transmitted through the heart also spread throughout the body and can be monitored and amplified with an instrument called an electrocardiograph.

Electrocardiogram is the recording of electrical activity of the heart.

P wave
- Atrial depolarization 0.1 second before atrial contraction
- Impulse spreads from SA to AV node.
- Widened may indicate: (1) Atrial enlargement (2) Mitral stenosis

19.6.2 PQ interval
- It is the time between beginning of atria contraction and beginning of ventricular contraction.
- Impulse travels from SA node to AV node, Av bundle, bundle branches, and finally to conduction myofibers.
- Prolonged PO interval may indicate AV damage or ischemia.

19.6.3 Q wave
- Deep and wide Q wave may indicate myocardial infarction.

19.6.4 R wave
- Increased size may indicate ventricular hypertrophy.

19.6.5 QRS complex
- Results from ventricular depolarization
- Precedes ventricular contraction
- Atria repolarize concurrently
- Time greater than 0.2 sec may indicate bundle branch block.

19.6.6 QT interval:

- Ventricles contract

19.6.7 ST interval

- Absolute refractory period
- The heart can't be forced to contract at this time.
- Time between end of depolarization of ventricles
- Plateau period
- Elevated levels may indicate M.I.
- Depressed levels may indicate ischemia/scarring or digitalis effect.

19.6.8 T wave

- Ventricles repolarize
- Relative refractory period
- Contraction at this point can yield ventricular fibrillation.
- Elevated when K levels are high
- Inverted when ischemia/scarring are present

19.7 Mechanical Events
19.7.1 The Cardiac Cycle

The cardiac cycle includes all events associated with the flow of blood through the heart during one complete heartbeat—that is, atrial systole and diastole followed by ventricular systole and diastole.

- Lasts about 0.8 sec
- Right and left sides act simultaneously, expelling the same volume.
- Chambers fill passively.
- Atria contract while ventricles relax.
- Ventricles contract while atria relax.
- Systole: contraction phase
- Diastole: relaxation phase

Phases

Relaxation period

1. Occurs at the end of heartbeat
2. All four chambers relaxed; pressure drops
3. Backflow of blood trapped by closing semilunar cusps of valves. This rebound produces the dicrotic notch.
4. All valves closed; isovolumetric relaxation
5. Ventricular pressure drops below atrial pressure, causing AV valves to open.

Ventricular filling

1. Rapid passive ventricular filling
2. Atrial systole—final 30 mL in 0.1 sec
3. End-diastolic volume (EDV): about 130 mL

19.7.4 Ventricular systole
1. AV valves shut—first heart sound
2. All valves closed—isovolumetric contraction
3. Ventricular pressure rises.
4. Semilunar valves open.
5. Ventricular ejection—0.5 sec
6. Semilunar valves close—second heart sound
7. Systole begins.
8. End-systolic volume (ESV): 60 mL

19.8 Cardiodynamics
1. Stroke volume = EDV – ESV
2. Cardiac output = Stroke volume x heart rate
3. End-diastolic volume (EDV): the amount of blood in each ventricle at the end of ventricular diastole (the start of ventricular systole)
4. End-systolic volume (ESV): the amount of blood remaining in each ventricle at the end of ventricular systole (the start of ventricular diastole)

19.8.1 Cardiac Output (CO)
Cardiac output is the amount of blood ejected from the ventricle into the aorta each minute. Cardiac output can be adjusted by changes in either heart rate or stroke volume.
- Cardiac output = Stroke volume x Heart rate (bpm)
- Cardiac output can be adjusted to meet demand.
  - (a) Heart rate can be adjusted by the activities of the autonomic nervous system or hormones.
  - (b) Stroke volume can be adjusted by changing the end-diastolic volume (how full the ventricles are when they start to contract).
  - (c) The end-systolic volume, i.e how much blood remains in ventricles in the ventricle after it contracts

19.9 Factors Affecting the Heart Rate
Autonomic innervation is located in the cardiovascular center of the medulla oblongata. It also involves input from the limbic system, proprioceptors, chemoceptors, and baroreceptors.

1. **Sympathetic stimulation** travels via the cardiac accelerator nerve that releases norepinephrine, which increases heart rate and calcium entry.
2. **Parasympathetic stimulation** travels via the vagus nerve by releasing acetylcholine. It slows down the pacemaker.
3. **Chemical: oxygen, pH and hormonal** release from the thyroid and adrenal medulla glands can affect the heart rate. Changes in Ca, Na, and K levels also affect the heart rate.

19.9.1 Factors Affecting the Stroke Volume
1. Preload: the degree of stretching experienced by ventricular muscle cells during ventricular diastole (this is based on the Frank-Starling law—more stretch results in stronger contraction. Preload depends on end diastolic volume(EDV), which varies with ventricular diastole length and venous pressure. Preload helps to balance right and left output.
2. Contractility: increase in contractile strength of the heart muscle.Several factors such as **glucagon, epinephrine, and digitalis increase contractility and calcium** inflow. This is referred to as a **positive ionotropic** effect. Factors such as anoxia, acidosis, and elevated

potassium level in the ECF decrease contractility (**negative ionotropic**).

1. Asystole: a condition in which the heart fails to contract.
2. Bradycardia: slow heart rate
3. Cardiac arrest: loss of effective heart beat; ventricular fibrillation
4. Cardiomegaly: cardiac hypertrophy
5. Cor pulmonale: a life-threatening condition of right-sided heart failure resulting from elevated blood pressure in the pulmonary circuit
6. Endocarditis: inflammation of the endocardium, usually confined to the endocardium of the heart valves
7. Heart palpitation: a heart beat that is unusually strong, fast, or irregular so that the person becomes aware of it; may be caused by drugs, emotional pressures, etc.

Review Questions

1. The blood returning to the heart from the systemic circuit will first enter the _____.

2. Blood returning to the heart from the pulmonary circuit first enters the _____.

3. The prominent muscular ridges that run along the surface of the auricles are called the _____.

4. The cusps of atrioventricular valves are attached to papillary muscles by a structure called the _____.

5. What is another name for visceral pericardium? _____

6. The relaxation phase of the cardiac cycle is called _____.

7. The left ventricle pumps blood to the _____.

8. Blood leaving the right ventricle enters the _____.

9. The pulmonary semilunar valves guards the entrance to the _____.

10. The first blood vessels to branch from the aorta are the _____ arteries.

11. The pacemaker cells of the heart are located in the _____.

12. List the sequence in which action potential would move through the conducting system of the heart. SA node to _____ to _____ to _____ to _____.

13. Depolarization of the ventricles is represented on an ECG by the _____.

14. The _____ is the amount of blood in a ventricle at the beginning of systole.

15. The first heart sound is heard when the _____.

16. What happens during the isovolumetric phase of ventricular systole? _____

_____

_____

17. The volume of blood ejected from each ventricle during contraction is called _____.

18. Define the "Starling's Law."

_____

19. Deep grooves and folds on the inner surface of the ventricles are called _____.

20. A slower-than-normal heart rate is called _____.

Chapter 20
Blood Vessels and Hemodynamics

20.1 Structure of Blood Vessel Walls
The walls of all blood vessels are composed of three distinct layers, or tunics. The tunics surround a central blood containing space, the lumen.

20.1.1 Tunica Interna
- Lined with endothelium containing a simple squamous epithelium resting on a basement membrane
- Internal elastic lamina containing elastic fibers that permit stretching and rebound

20.1.2 Tunica Media
- Contains circularly arranged smooth muscle cells and sheets of elastin
- Its activity is regulated by vasomotor nerve fibers of the sympathetic nervous system (SNS), permitting changes in diameter through vasoconstriction and vasodilation.

20.1.3 Tunica Externa (Adventitia)
- Contains loosely woven collagen fibers for flexible support
- Infiltrated with nerve fibers and lymphatic vessels
- Network of elastin fibers in larger veins
- Vasa vasorum (network of small blood) vessels nourish the wall.

20.2 Arterial System
Arteries are vessels that carry blood from cardiac ventricles to another area of the body. They do not have valves.

20.2.1 Types of Arteries

1. **Elastic (conducting)** are large arteries with thin walls compared to lumen diameter. They carry large volumes of blood quickly. Their walls are more elastic but less muscular. Examples are the aorta, subclavian artery, and common iliac artery.

2. **Muscular** (distributing) arteries are intermediate in size and have thicker walls compared to lumen diameter. They help to adjust the rate of blood flow. Their walls are less elastic but more muscular.

3. **Arterioles** have lumen diameters ranging from 0.3 mm down to 10um. Arterioles are the major sites of blood pressure regulation. Smooth muscle controls the peripheral resistance; vasoconstriction raises BP, and vasodilation lowers BP.

4. **Metarterioles** provide a direct route through capillary beds. They help to regulate blood flow.

5. **Capillaries** are sites of exchange between cells and the blood. Blood flow to the capillary bed is regulated by metarterioles as the precapillary sphincter open and close.

Note: Bloodflow through the capillary network is slow and intermittent, rather than steady. This is referred to as vasomotion; it is 5–10 times/minute.

20.3 Types of Capillaries
Continuous capillaries are most abundant in the skin and muscles. The endothelial cells provide uninterrupted lining, and adjacent cells are joined together by tight junctions. Some **intercellular clefts** exist where the membranes are not joined together. Continuous capillaries constitute the structural basis

of the blood-brain-barrier (BBB).

Fenestrated capillaries have endothelial cells riddled with oval pores or **fenestration.** They are found where active capillary absorption or filtrate formation occurs—e.g., the small intestine, endocrine organs, and the kidneys.

Sinusoids are modified leaky capillaries found in the liver, bone marrow, lymphoid tissue, and some endocrine organs. They contain larger intercellular clefts allowing larger molecules and blood cells to pass. Kupffer cell are macrophages that form part of the discontinuous endothelium in the liver

Vascular anastomoses are found where vascular channels unite. They provide **direct** connection of two or more blood vessels serving an area. They may occur between artery to artery, vein to vein, or arteriole to venule, where they provide collateral (alternate) pathways. They are common in abdominal organs, the brain, and the heart. Damage to anastomoses yields necrosis of the area served.

Note: Arteriovenous anastomoses act as thoroughfare channels that shunt capillary beds that connect arterioles and venules.

20.4 The Venous System

Blood is carried from the capillary beds toward the heart by veins. En route, the venous vessels increase in diameter, and the walls gradually thicken as they progress from venules to the larger and larger veins.

Venules range from 8 to 100 μm in diameter. They are extremely porous, allowing fluid and WBC to move easily from the bloodstream through their walls.

Veins: Venules join to form veins. Veins have thin tunica interna, tunica media, and thick tunica externa. Veins are not as strong as arteries, and they have valves to prevent back flow of blood. They are found in veins of limbs and other areas if needed to fight gravity.

Note: Sixty-five percent of the body's total blood supply is found in the veins at any time.

20.5 Velocity of Blood Flow

a. Blood velocity is slower in vessels with larger cross-sectional areas.

b. Blood velocity is faster in vessels with smaller total cross-sectional areas.

c. Circulation time is one minute at rest. This is the time it takes for blood to pass through the heart, lungs, and back to the heart.

20.5.1 Volume of Blood flow

1. **Cardiac output:** Stroke volume x heart rate
2. Blood volume affects cardiac output and blood pressure. The normal value should be five liters.
3. **Blood pressure (BP)**
   * pressure of blood against vessel wall.
   * Contracted ventricles/relaxed ventricles
   * Systolic/diastolic == 120/80
4. **Peripheral resistance:** friction that slows blood flow; caused by three factors:
   (1) **Blood viscosity:** depends on packed cell volume, protein concentration, and degree of hydration
   (2) **Blood vessel length:** the longer the length, the greater the resistance; shorter length, less resistance
   (3) **Blood vessel** radius: $R \propto 1/r^4$; smaller radius, greater resistance; larger

radius, less resistance

Note: Systemic vascular resistance is controlled via the vasomotor center of the medulla oblongata.

20.6 Blood Flow through Capillaries and Capillary Dynamics

Diffusion is the most important method of exchange. Lipid-soluble substances cross plasma membranes, and water-soluble substances pass through fenestrations and intercellular clefts.

Vesicular transport is used for the transport of large, nonlipid-soluble molecules. These materials are moved in by the process of endocytosis.

Bulk flow is important in balancing blood and interstitial volumes. It is a passive process involving mass migration of ions, molecules, and particles in the same direction. Forces causing the movement include air pressure and hydrostatic pressure.

20.7 Hydrostatic pressure
1. Blood hydrostatic pressure (BHP) pushes fluid out of capillaries.
2. BHP value is typically 30 mm Hg at the arterial end and 16 mm Hg at the venous end.
3. Interstitial fluid hydrostatic pressure (IFHP) is the pressure of interstitial fluid within the tissue spaces. It is -3 mmHg because of the suction effect produced by the lymphatic vessels.
4. Net hydrostatic pressure = BHP – IFHP
$$= 30 - (-3) = 33 \text{ mmHg}$$

20.7.1 Osmotic pressures (oncotic pressure)
1. Blood colloid oncotic pressure (BCOP) is the difference in osmotic pressure between the blood and the interstitial fluid. This is caused by large plasma proteins that pull water from capillaries. BCOP is 28 mm Hg.
2. Interstitial fluid osmotic pressure (ICOP) is about 8 mm Hg.
3. Net osmotic pressure = BCOP – ICOP = 28 – 8
$$= 20 \text{ mm Hg}$$
4. Pulls water from capillaries

20.7.2 Net filtration pressure (NFP)
1. The value of NFP determines the direction of fluid movement.
2. NFP = Net hydrostatic pressure – Net osmotic pressure
$$= 33 - 20$$
$$= 13 \text{ mm Hg}$$

3. Under normal condition, net filtration pressure (NFP) is positive at the arterial end, encouraging fluid to move out.
4. NFP is negative at the venous end, encouraging fluid to move in.
5. The balance between filtration and reabsorption depends on tissue function.

20.8 Venous return is the volume of blood flowing back to the heart. It depends upon the following factors:
1. The pressure at the right atrium must be 0 mm HG.
2. Veins must have low resistance.
3. The valves must be fully functional.
4. Skeletal muscle contractions open valves and move blood.
5. Diaphragm contraction changes pressures.

20.9 Control of Blood Pressure and Blood Flow

The principal parameters that are involved with the adjustment of blood pressure and distribution are under the direction of a negative feedback system. They are as follows:

20.9.1 Control of Blood Pressure by Actions on the Heart

The **cardiovascular center** is located in the medulla oblongata.Output from the cardiovascular center flows along sympathetic (cardiac stimulator center)and parasympathetic (cardiac inhibiting center) fibers

    a. Sympathetic impulses along cardiacaccelerator nerves increase heart rate by releasing epinephrine or norepinephrine.

    b. Parasympathetic impulses along vagus (X) nerves decrease heart rate by releasing acetylcholine.

    c. The sympathetic division also causes vasoconstriction or vasodilation as it continuously sends impulses to smooth muscle in blood vessel walls via vasomotor nerves.

20.9.2 Neural Regulation of Blood Pressure

Baroreceptors are found in blood vessel walls and the right atrium, where they function to protect circulation against short-term changes in blood pressure as posture changes.

1. The carotid sinus reflex protects the blood supply to the brain.
2. The aortic reflex maintains adequate blood pressure in the systemic circuit as a whole.
3. Carotid and aortic bodies transmit impulses to the cardioacceleratory center when blood oxygen and pH drop sharply or carbon dioxide levels rise, causing cardiac output to increase.

20.9.3 Hormonal regulation of BP provides short-term and long-term regulation of cardiovascular performance.

1. **Antidiuretic hormone** is produced in the hypothalamus and released in the neurohypophysis. It causes vasoconstriction in association with hemorrhages. ADH is inhibited by alcohol.
2. **Angiotensin II** appears in the blood after the release of the enzyme renin by juxtaglomerular cells in the kidneys in response to low renal blood pressure. It causes vasoconstriction and stimulates aldosterone secretion, in turn increasing absorption of sodium ions and water.
3. **Natriuretic peptide** is released by cells in the cardiac atria. It causes vasodilation and promotes loss of sodium ions and water.
4. **Histamine** is released from mast cells. It acts as a vasodilator.
5. **Kinins** are found in the plasma. They are vasodilators.

20.9.4 Autoregulation of BP refers to local, automatic adjustments of blood flow in a given region to match the particular needs of the tissue.

20.9.5 Chemical Mediators

1. changes in oxygen and carbon dioxide levels
2. tissue stretching
3. release of vasodilators, such as nitric oxide, lactic acid, K and H ions, and adenosine
4. release of vasoconstrictors, such as eicosanoids, thromboxane A2, angiotensins, and endothelins

20.13 Physical Changes

- Warming causes vasodilation.
- Cooling causes vasoconstriction.

Review Questions

1. The muscular layer of a blood vessel is the _____.
2. Capillaries that have a complete lining are called _____.
3. The function of the precapillary sphincter is to _____.
4. What condition would cause blood flow to a tissue to increase? _____
5. The difference between the systolic and diastolic pressures is called the _____ pressure.
6. What is the effect of renin released from the juxtaglomerular cells? _____
_____
_____
7. An important artery that supplies blood to the brain is the _____ artery.
8. Blood from the brain returns to the heart by way of the _____ vein.
9. After passing through the rib, the subclavian artery becomes the _____ artery.
10. The internal carotids and the basilar artery are interconnected by an anastomosis called the _____.
11. The _____ are small blood vessels that carry blood to the tunica media of large arteries and veins.
12. Define the term vasomotion. _____
13. _____ arteries are a group of arteries that service a single capillary network.
14. What is metarteriole? _____
15. The blood vessel that carries blood to the arm and shoulder is the _____.
16. The blood vessel that supplies blood to the head and neck is the _____.
17. The blood vessel that supplies blood to the liver, stomach, and spleen is the _____.
18. The blood vessel that supplies blood directly to the liver is the _____.
19. The small intestine and most of the large intestine receive blood from the _____.
20. Ovaries or testes receive blood supply from the _____.
21. The kidneys receive blood from the _____.
22. Blood from the inside of the cranium is drained by the _____.
23. The _____ receives blood from the diaphragm.
24. _____ drain fluid from the interstitial space back to the general circulation.
25. A(n) _____ is a direct connection between an arteriole and a venule.
26. The layer of the arteriole wall that provides the properties of contractility and elasticity is the _____.
27. Venoconstriction _____ the amount of blood within the venous system, which _____ the volume in the arterial system.
28. The femoral artery is an example of a(n) _____ artery
29. The two common iliac veins form the _____.
30. Nutrients from the digestive tract enter the _____ vein.

Chapter 21
The Lymphatic System and Immunity

21.1 The lymphatic system consists of two semi-independent parts: (1) a network of lymphatic vessels and (2) various lymphoid tissues and organs scattered throughout the body.

21.1. Lymphatic vessels
1. Return leaked fluid and plasma proteins to systemic circulation
2. Form a one-way system through which lymph flows only toward the heart
- Present everywhere except bones and teeth, bone marrow, and entire nervous system
- Lymphatic capillaries (lacteals) → collecting vessels → trunks → ducts

Lymphatic trunks are formed by the union of the largest collecting vessels: paired lumbar, bronchiomediastinal, subclavian, jugular, and intestinal
- Lymph eventually drains into one of two large ducts in the thoracic region:
- **Right lymphatic duct** drains lymph from the right upper arms and the right side of the head and thorax.
- **Thoracic duct** receives lymph from the rest of the body. It arises from cisterna chyli, which collects lymph from the two large lumbar trunks that drain the lower limbs and intestinal trunk from the digestive system.

21.2 Lymph Organs
1. **Lymph nodes** are located in the superficial cervical, axillary, and inguinal regions of the body. They are also found deep in the GI tract. They contain macrophages (which filter blood) and lymphocytes (which provide immune response).
2. **Tonsils:** There are three sets of tonsils. They are located in the pharynx region. They are the **pharyngeal tonsil, palatine tonsils,** and **lingual tonsil.** They make lymphocytes that remove pathogens from the throat area.
3. **Peyer's patches** are located in the ileum. They produce macrophages that capture and destroy bacteria.
4. The **spleen** is located posterior to the stomach. It contains red and white pulp. The **red pulp** removes worn-out and defective blood cells. It stores the breakdown products. The **white pulp** is a lymphatic tissue that filters debris and pathogens from blood. It is also the site of lymphocyte production.
5. The **thymus** is located in the mediastinum. It activates T cells and also secretes thymosin and other hormones.
6. **Bone marrow** is the site lymphocyte production.

21.3 Innate and Adaptive Body Defenses

Innate (nonspecific) defenses do not require previous exposure to pathogens. They respond immediately to infections—e.g., skin, saliva, vomiting, and fever. They may provide mechanical barriers or chemical barriers.

Mechanical
1. Skin: The epidermis and hair provide barriers against pathogens.
2. Mucous membrane: It lines all cavities that open to the exterior. It produces mucous onto hair and cilia, where pathogens are trapped.
3. Epiglottis: It prevents entry of objects into the lower respiratory tract.
4. Lacrimal apparatus: It makes and drains tears and washes the eye surface.
5. Salivary glands: They produce saliva, which dilutes pathogens and washes the oral

cavity.
6. Urine: The flow of urine cleanses the urethra.
7. Vagina secretion: It cleanses the vagina.
8. Defecation: It eliminates pathogens.
9. Emesis: It expels pathogens.

Chemical Barriers to Disease:
1. Sebum: unsaturated fatty acids that inhibit fungal and bacterial growth
2. Perspiration: It washes pathogens away.
3. Lysozyme: It digests bacterial walls. It is found in tears, saliva, sweat, nasal secretions, and tissue fluid.
4. Hyaluronic acid: slows the spread of localized infections and toxins
5. Acid: destroys bacteria and toxins. It is found in skin, gastric juice, and vaginal secretion.

21.3 Antimicrobial proteins enhance innate defense by attacking microorganisms directly or hindering their ability to reproduce. They act as the second line of defense, used when pathogens pass through skin and mucous membranes. They are found in the blood and interstitial fluid.

1. **Transferrins** are iron-binding proteins that reduce bacterial growth.
2. **Interferons (IFNs)** are proteins produced when a viral infection is present. They are produced by lymphocytes, macrophages, and fibroblasts. They cause uninfected cells to block viral reproduction by producing PKR protein. Types I (α, β) inhibit cell growth and suppress tumor formation, type II (γ) enhances phagocytosis and natural kill cell activity.
3. **Complement system** is a system of about 20 proteins found in the plasma and cell membrane. The proteins include C1 through C9 (C1 is actually three proteins), factors B, D, and P (properdin), plus several regulatory **proteins. Protein C3 is the key protein that activates the two pathways (classical and alternative pathways).**

21.5 Natural Killer Cells
In addition to B cells and T cells, natural killer (NK) cells have the ability to kill a wide variety of microbes and tumors.

21.6 Phagocytosis
The third line of defense is phagocytosis, the ingestion and destruction of microbes. Phagocytes participate in both nonspecific defenses and immunity. There are two categories of phagocytes: granulocytes (neutrophils, osinophils, and basophils) and macrophages (fixed and wandering). There are five phases of phagocytosis:

1. **Chemotaxis:** Phagocytes move to the infected area by the process of diapedesis.
2. **Adherence:** The phagocytes adhere to particles' rough surfaces (opsonization).
3. **Ingestion:** The phagocyte engulfs the particles by the process of endocytosis into a phagocytic vesicle.
4. **Killing:** The phagocytic vesicles fuse with lysosomes, and the pathogens are destroyed.
5. **Expulsion:** Undigested materials are expelled by the process of endocytosis.

Cytolysis
Membrane attack complex (MAC): perforate cell and ensure lysis by interfering with cell's ability to eject Ca^{2+}.

21.7 Inflammation
Inflammation is a response to tissue injury. Inflammatory response is a *good* thing!

1. Symptoms: heat, pain, redness, and swelling
2. Vasodilation and increased vessel permeability allow more blood to flow through the area and flush out debris and pathogens. They bring protective substances to area.
3. Mast cells release histamin, heparin, and prostaglandins.
4. Clotting factors and complement proteins enter area.
5. Local temperature elevation increases the activity of enzymes and phagocytes.
6. Neutrophils attack debris and bacteria (squeezing through capillaries—diapedesis).
7. Macrophages engulf pathogens and debris.
8. Cytokines released by phagocytes stimulate tissue repair.
9. Tissue damage and chemicals stimulate pain receptors.
10. Specific defenses are activated.

Fever is an abnormally high temperature caused by bacteria, toxins, and viruses. It is mediated by interleukin and prostaglandin. Fever creates a hostile environment for pathogens.

21.8 Adaptive (Specific) Defenses

There are three important aspects of the adaptive immune response: (1) antigenicity, (2) memory, and (3) specificity. Immune cells originate from the bone marrow, where they also develop antigen receptors. B Cells mature in bone marrow, while T cells mature in the thymus gland and differentiate into CD4+ and CD8+. This process is referred to as immunocompetence (the ability to carry out immune responses if properly stimulated).

21.9   Types of Immune Response
Cell-mediated Immune Response (CMI)

- This is cellular immunity where cells attack cells.
- CD8+ cells becomes killer T cells and engage in a direct attack on an antigen.
- CMI works best against intracellular pathogens such as fungi, parasites, viruses, and cancer cells.
- It also attacks foreign tissue transplants.
- CD4+ cells become helper T cells that aid both CMI and AMI.

Antibody-mediated Immune Response (AMI)

This is humoral immunity provided by antibodies present in the body's "humor" or fluids. B cells produce antibodies, and antibodies inactivate antigens. AMI works best against antigens in body fluids and extracellular pathogens.

21.10 Antigens

Antigens are substances that can mobilize the immune system and provoke an immune response.
Characteristics
1. They stimulate immune response.
2. They react specifically with an antibody.
3. They are either proteins, such as nucleoproteins, lipoproteins, and glycoproteins, or polysaccharides.
4. They provoke response from T cells and B cells.

Haptens or incomplete antigens: Some small molecules, such as hormones, are not immunogenic until they link up with the body's own proteins (allergies).
Epitopes: Only certain parts of an antigen are immunogenic—the antigenic determinants that trigger immune responses.
Antigen receptors exhibit great diversity due to genetic recombination.

21.11 Self-antigens: HLA Proteins

Human leukocyte–associated antigens (HLA) are unique to each individual. Our immune system has been programmed to recognize them as self-antigens. These self-antigens aid in the detection of foreign invaders.

Type I: They are found on cell membranes except red blood cells.

Type II: They are found in thymus cells, activated T cells, and antigen-presenting cells.

21.12 Cells of the Adaptive Immune System

1. **Antigen-presenting cells (APCs)** do not respond to specific antigens, but they play an axillary role.

2. **Lymphocytes or B cells** oversee humoral immunity—i.e., they can recognize and bind to antigens in the extracellular fluid.

3. **T lymphocytes or T cells** are non-antibody–producing lymphocytes that constitute the cell-mediated arm of immunity. They can only recognize fragments of antigenic proteins that first have been processed and presented in association with MHC self-antigens.

Lymphocytes originate in red bone marrow from hemopoietic cells (hemocytoblasts). Whether a given lymphocyte will mature into B cell or T cell depends on where in the body it becomes **immunocompetent**. T cells become immunocompetent in the thymus and B cells in the bone marrow.

21.13 Antigen-presenting cells (APC)

engulf foreign particles and present fragments of these antigens like flags on their surfaces (antigen presenters) so that T cells can recognize them. Examples of APCs are dendritic cells such as interstitial cells in connective tissues, Langerhans cells of the epidermis of the skin, macrophages, and activated B lymphocytes.

Steps in processing and presenting an exogenous antigen by an APC:

1. Phagocytosis or andocytosis of the antigen
2. Partial digestion of the antigen
3. The fragments of digestion are enclosed in vesicles.
4. The peptide fragments bind to MHC-II molecules.
5. They are released by the process of exocytosis.

21.14 Cytokines

are small protein hormones needed for many normal cell functions, such as regulation of immune responses. Examples of cytokins are lymphokines and monokine.

21.15 Antibodies

are called immunoglobulins or Igs. They are soluble proteins secreted by activated B cells or plasma cells in response to an antigen. Antibodies specifically bind to antigens.

21.15.1 Basic Antibody Structure

An antibody constitutes the gamma globulin part of blood protein with a glycoprotein side chain.

Composition: A typical antibody has the following composition.

- Two identical heavy (H) chains (contains approx. 400 amino acids each)
- Two identical light (L) chains (half as long)
- Each light chain is attached to a heavy chain by a disulfide bond.
- The ends of the combined heavy and light chains form the variable region of the antibody.
- The variable (V) region is the part that combines with the antigenic determinant of the antigen.
- Different antibodies have different variable regions and are specific for different antigens.
- The constant (C) region determine the five classes of antibodies: IgM, IgA, IgD, IgG, and IgE (MADGE).

21.15.2 Actions of Antibodies

Antibodies act in two ways: direct attack and complement activation.

Direct attack

   a. Agglutination: occurs when cell-bound antigens are cross-linked. IgM is a potent agglutinating agent.

   b. Precipitation: Soluble molecules are cross-linked into large complexes that settle out of solutions, making them easily engulfed by phagocytes.

   c. Neutralization: The antibody masks dangerous parts of bacteria exotoxins and viruses.

   d. Activation of a complement through the C1 protein promotes lysis.

Activation of complement: When the variable region of an antibody combines with an antigen, the constant region can activate the complement cascade through the classical pathway. The destruction of the cell membrane of the target cell stimulates an inflammatory response. This attracts neutrophils, monocytes, macrophages, and eosinophils to the sites of infection and kills bacteria by lysis.

Note: As elaborate as the complement system may appear, it has two main goals:

   1. Stimulation of inflammatory response

   2. Cell lysis and bacterial destruction

21.15.3 Actions of Specific Antibodies

   a. IgM activates the complement and acts as an antigen-binding receptor on the surface of B cells. It is responsible for transfusion reactions in the ABO blood system.

   b. IgA is found in body secretions such as saliva, sweat, intestinal juice, and milk, and helps prevent attachment of pathogens to epithelial cell surfaces.

   c. IGD functions as antigen-binding receptors on B cells. It is important in B cell activation.

   d. IgG is the most abundant and diverse antibody in the plasma. It protects against bacteria, viruses and toxins circulating in blood and lymph. It readily fixes complement. It crosses the placenta to confer passive immunity to the fetus. It is responsible for Rh reactions.

   e. IgE are released in during allergic reactions. They attach to basophils and mast cells to stimulate inflammatory reactions.

21.16 Self-Tolerance

It is the body's ability to recognize its own MHC molecules (immunological tolerance). The B cells and T cells do not react to the body's own proteins (antigens). Processes:

   + selection: Lymphocytes that interact weakly with self-MHC molecules are allowed to continue to develop in the thymic cortex.

   2. **–selection** gets rid of cells with T receptors that recognize self-proteins (occurs in thymic medulla)

   3. **Anergy:** inactivation of some self-reactive B and T cells

   4. **Deletion**: Self-reactive T cells are killed outright.

Immunology and Cancer

   a. Tumor cells display cell-surface antigens (SAs).

   b. Tumors caused by viruses are recognized by killer T cells, macrophages, and natural killer cells

   c. Surface **antigen**: The monoclonal antibody reaction is used in diagnosis of some tumors. An example is the PSA test for prostate cancer.

21.17 Homeostatic Imbalances of Immunity

Under certain circumstances the immune system is depressed or even fails, or it exhibits activities that damage the body itself.

21.17.1 Immunodeficiencies can be congenital or acquired. Examples are as follows:

a. The production or functions of immune cells, phagocytes, or complements is abnormal.
b. Severe combined immunodeficiency (SCID) syndromes is a congenital condition.
c. Children afflicted with SCID have little or no protection against disease-causing organisms.
d. Hodgkin's disease (cancer of the lymph node) is an acquired condition.
e. Acquired immune deficiency syndrome (AIDS) interferes with the activity of helper T (CD4) cells.
f. Autoimmune disease

21.18 Self-tolerance breaks down (positive or negative selection fails, and anergy is reversed). Examples are:

1. Multiple sclerosis (MS): White matter of the brain and spinal cord are destroyed.
2. Myasthenia gravis: Communication between nerves and skeletal muscles is impaired.
3. Grave's disease: excessive amount of thryoxine production by the thyroid gland
4. Type I (juvenile) diabetes mellitus: Pancreatic beta cells are destroyed, resulting in deficit insulin production.
5. Systemic lupus erythematosus (SLE): a systemic disease that affect kidneys, the heart, lungs, and skin
6. Glomerulonephritis: a severe impairment of renal function
7. Rheumatoid arthritis (RA): systematically destroys joints

21.19 Hypersensitivities

Hypersensitivity is associated with elevated sensitivity to antigens normally tolerated by most people. It is also known as allergic reactions or allergies.

Allergens: antigens that induce allergic reactions, such as pollen or animal dander.

Reaction Types

Type I (Anaphylaxis)
- begins within seconds and is most common
- requires previous antigen exposure (sensitizes the person)
- Steps are mediated by interleuken 4 secreted by T cells.
- Interleuken 4 stimulates B cells to mature into IgE antibodies.
- IgE binds to the surfaces of mast cells and basophils.
- Histamine is released, stimulating inflammatory responses.

Cytotoxic (type II) reactions occur when antibodies bind to antigens on specific body cells and stimulate phagocytosis and lysis of the antigens. The process is stimulated in cases of incompatible transfusions. It involves IgG and IgM antibodies.

Immune complex (type III) hypersensitivity occurs when antigens are widely distributed throughout the body or blood-forming insoluble antigen-antibody complexes that cannot be cleared from a particular area.

Delayed hypersensitivity (type IV) reactions is a cell-mediated reaction that is slower than antibody-mediated hypersensitivity. Cytotoxic cells attack specific antigens. It is often triggered by intracellular bacteria and poison ivy toxin.

21.20 Immunological Memory
- It is due to the presence of long-lived antibodies and very long-lived lymphocytes that arise during the proliferation and differentiation of antigen-stimulated B and T cells.
- Immunization against certain microbes is possible because memory B cells and memory T cells

offer the primary response to an antigen.

21.20.1 Active and Passive Humoral Immunity
Naturally Acquired

(1) Active: Infection requires contact with pathogen.
(2) Passive: Antibodies pass from mother to fetus via placenta or to infant in mother's milk.

Artificially Acquired

(1) Active: Dead vaccine or attenuated pathogens given to stimulate antibody production
(2) Passive: Injection of immune serum (gamma globulin)

Functions of Innate Immunity Cells

- Neutrophils are phagocytes produced in the red bone marrow. They play a major role in inflammation and phagocytosis.
- Eosinophils are produced in the red bone marrow and enter the tissues where they break down chemicals released by basophils during inflammation reaction, thus reducing inflammatory response.
- Basophils release chemicals that promote inflammation.
- Mast cells are found in connective tissues where they promote inflammation by releasing histamine.
- Monocytes leave the blood vessels through the process of diapedesis and enter tissues where they become macrophages.
- Macrophages are monocytes that leave the blood, enter tissues and become highly phagocytic.
- Natural killer cells are lymphocytes produced in the bone marrow. They lysis tumor cells and virus-infected cells.

Functions of Adaptive Immunity Cells.

- B cells differentiate in the bone marrow.to become plasmacells or memory cells.
- Plasma cells proliferate to become antibodies that are capable of destroying antigens.
- Memory B cells are cells that remain in the plasma after immune system reaction. They react quickly to subsequent exposure to similar antigen.
- Cytotoxic T cells are capable of direct attack of cells by lysis or by producing cytokines.
- Helper T Cells activate B cells and effector T cells
- Suppressor T cells inhibit B cells and effector T cells.
- Dendritic cells help to process antigen and B cells and T cells activation

Review Questions

1. **Most of the lymph returns to the venous circulation by way of the _____.**
2. **Where are Peyer's patches located?**

   _____

3. **Areas of the spleen that contains large numbers of lymphocytes are known as _____.**

4. **Areas of the spleen called "red pulp" contain _____.**

5. Which events trigger an inflammatory response?

_____

6. The release of interleukin-1 by macrophages causes _____.
7. Various macrophages are derived from _____.
8. What event leads to the classic pathway of complement fixation? _____

_____

9. What is innate immunity?

_____

10. Immunity that results from the natural exposure to an antigen in the environment is called _____.
11. The cells responsible for the production of circulating antibodies are _____ cells.
12. The cells responsible for cellular immunity are the _____ cells.
13. Lymphocytes that attack foreign cells or body cells infected with viruses are _____ cells.
14. What is the function suppressor T cells?

_____

15. What is the function of NK cells?

_____

16. Stem cells that will form T cells are modified in the _____.
17. When an antigen is bound to a class I HLA molecule, it can stimulate a _____ cell.
18. Class II molecules are only attached to _____ and _____.
19. An antigen that binds to a class II HLA can stimulate a _____ cell.
20. When does class II MHC appear in the cell membrane?

_____

21. _____ are fixed macrophages located in the epithelia of the skin.
22. The fixed macrophages found in the liver are called _____.
23. _____ are antibodies that make it easier for phagocytes to engulf their target cells.
24. What is the role of properdin in the process of complement system? _____

_____

25. _____ are antibodies found in body fluids.
26. What is opsonization?

_____

Chapter 22
The Respiratory System

The major function of the respiratory system is to supply the body with oxygen and remove carbon dioxide formed as a result of metabolic activities. There are four distinct processes, collectively called respiration.

1. **Pulmonary respiration (ventilation)** is the movement of air into and out of the lungs— breathing— which involves inspiration and expiration.
2. **External respiration** is the gas exchange (oxygen loading and carbon dioxide unloading) between the blood and lungs.
3. **Internal respiration** is the exchange of gases between blood and cells.
4. **Cellular respiration** is the reaction of glucose with oxygen to form carbon dioxide, water, and energy in the form of ATP.

22.1 Structures of the Respiratory System

The respiratory system consists of the nose, pharynx, larynx, trachea, bronchi, and lungs.

1. **Nose**
   a. The external portion of the nose is made of cartilage and skin and is lined with mucus membranes. The external nares (nostrils) open to the exterior and are separated by the nasal septum.
   b. The internal portion communicates with the paranasal sinuses and nasopharynx through the internal nares (choanae). The paranasal sinuses are located in the frontal, sphenoidal, ethmoidal, and maxillary bones.
   c. The anterior portion of the nasal cavity is called the vestibule.
   d. The function of the nose is to warm, moisten, and filter the incoming air. It also receives olfactory stimuli. It also plays a role in resonance and speech modification.

2. **Pharynx (Throat)**
   a. The pharynx extends from the internal nares to the cricoid cartilage (laynx).
   b. It is a muscular tube lined by a mucous membrane.
   c. The throat is made up of the following anatomical regions: nasopharynx, oropharynx, and laryngopharynx.
   d. The nasopharynx functions in respiration. The oropharynx and laryngopharynx function in digestion and in respiration.

3. **Larynx**
   a. The larynx joins the pharynx to the trachea.
   b. It contains the thyroid cartilage (Adam's apple); the epiglottis; the cricoid cartilage; and the arytenoid , corniculate, and cuneiform cartilages. The epiglottis protects the airway by covering the glottis.
   c. The larynx contains the vocal cords, which produce sound.

22.1.2 The Trachea

- The trachea is also known as the windpipe.
- It leads from the larynx into the bronchial tree.
- The C-shaped cartilage holds the trachea open and allows for expansion.
- It consists of several layers: mucosa, submucosa, and adventitia.
- The mucosa have goblet cell–containing psudostratified epithelium.

The Bronchi and Subdivisions
22.1.3 The Bronchial Tree

- The right and left primary bronchi are formed by the division of the trachea.
- The diameter decreases as branching increases.
- Each primary bronchus serves a lung. The right is wider, shorter, and more vertical.
- The left bronchus is smaller in diameter, longer, and more horizontal.
- The secondary bronchi serve a lobe: three on the right; two on the left.
- Tertiary (segmental) bronchus serves the bronchopulmonary segment.
- Bronchioles: Terminals extend into alveolar clusters.
- Respiratory portions extend directly to alveoli. The sympathetic nervous system dilates the bronchioles, and the parasympathetic nervous system constricts them.

22.2 The Lungs and Pleural Coverings

The lung is enclosed by pleurae. The parietal pleura cover the thoracic wall and superior surface of the diaphragm. The visceral covers the external lung surface. The pleura produce pleural fluid that fills the pleural cavity between them.

22.2.1 Gross Anatomy of the Lungs

- Base: fits on the diaphragm
- Apex: extends into root of neck
- Costal surface: lies against ribs
- Mediastinal surface: faces the heart, which contains the hilum that serves as entrance/exit for blood vessels, bronchi, and nerves
- The right lung has three lobes (superior, middle, inferior) and has two fissures (oblique and horizontal).
- The left lung has two lobes (superior and inferior), one fissure (oblique), and the cardiac notch.

22.2.2 Histology of the Lung

Each lung can be subdivided as follows

- Lungs
- Segment
- Lobule
- Alveoli
- **Alveolus**

Internal Structure

- The walls are composed of a single layer of simple squamous cells called Type I cells.
- Type II cells produce surfactant.
- Alveoli pores connect adjacent alveoli, serving to equalize air pressure in the lung or providing alternate air routes in the alveoli.
- Alveoli macrophages, monocytes, and fibroblasts are also present.

Note: The alveolar-capillary membrane ($0.5\mu m$ thick) serves as the respiratory membrane. It is composed of:

- alveolar wall
- epithelial basement membrane
- capillary basement membrane
- capillary endothelial cells

22.3 Pulmonary Ventilation (Respiration)

Respiration is the exchange of gases between the atmosphere and lungs.

Normal inspiration (inhalation) occurs when intrapleural pressure decreases and air rushes into the lungs. (Boyle's Law)

Forced inspiration occurs when the body needs more air exchange. Additional muscles are used to raise the thoracic cage, namely the sternocleidomastoid, scalene, and pectoralis major muscles.

Normal expiration is a passive process that depends on the natural elasticity of the lungs as the diaphragm relaxes; intrapleural pressure increases, and air is pushed out of the lungs. The thoracic volume decreases as the pressure rises (Boyle's Law).

Forced expiration is an active process produced by the abdominal muscles and internal intercostalis when the body needs more air exchange.

The following factors affect ease of respiration:
1. Compliance: lung elasticity and surface tension, surfactant
2. Airway resistance

22.4 Respiratory Volumes and Pulmonary Function Tests

The amount of air that is flushed in and out of the lungs varies substantially depending on the condition of inspiration and expiration.

- 1 respiration = 1 inspiration + 1 expiration
- Normal value is about 12 per minute or six liters per minute.
- The apparatus that is used to measure respiratory volumes is called a spirometer.

Pulmonary Volumes (Specific Conditions)

Tidal volume (V): amount of air that moves into and out of the lungs with each breath (ca 500 ml)

2. Inspiratory reserve volume (IRV): amount of air that cn be inspired forcibly beyond the tidal volume

3. Expiratory reserve volume (ERV): amount of air that can be evacuated from the lungs after a tidal expiration (about 1000 to 1200 ml)

4. Residual volume (RV): amount of air that remains in the lungs even after a strenuous expiration (about 1200 ml). This amount of air keeps the lung patent and prevents collapse.

5. Minimal volume (MV): amount of air that would remain in the lungs if they were allowed to collapse (about 30 to 120 ml)

22.5 Pulmonary Capacities (Combined Conditions)

Respiratory capacities can be calculated by adding the values of various volumes.
1. **Inspiratory capacity**: amount of air you can draw into your lungs after you have completed a respiratory cycle: (TV + IRV)
2. **Functional residual capacity**: amount of air remaining in your lungs after you have completed a quiet respiratory cycle (RV + ERV)
3. **Vital capacity:** the maximum amount of air that you can move into and out of your lungs in a single respiratory cycle (IRV + TV + ERV)
4. **Total lung capacity**: total volume of your lungs (TV + IRV + ERV + RV + MV)

22.6 Transport of Respiratory Gases in the Blood

1. Molecular oxygen is carried in the blood in two ways: bound to hemoglobin within the red blood cells and dissolved in the plasma. Most of the oxygen is bound to hemoglobin (Hb):

   $Hb + O_2 \leftrightarrow HbO_2$

         (Oxyhemoglobin)

         Lungs (loading)

$$Hb + O_2 \leftrightarrow HbO_2 + H^+$$
Tissues
(unloading)

2. Carbon dioxide dissolves into small amounts in plasma but it also binds to hemoglobin: $Hb + CO_2 \leftrightarrow HbCO_2$ (Carbaminohemoglobin)
Carbon dioxide is also transported as bicarbonate
ions: $CO_2 + H_2O \leftrightarrow H_2CO_3$
$$H2CO_3 \leftrightarrow H + HCO_3$$

22.6.1 Behavior of Gases

The behavior of gases determines the amount of gas present to form a respiratory gradient.
1. **Boyle's law:** The pressure of a gas is inversely proportional to its volume when temperature is kept constant; i.e., greater pressure leads to smaller volume.
2. **Dalton's law of partial pressures**: The total pressure exerted by a mixture of gases is the sum of the individual gas pressures.
3. **Charles's law**: Higher gas temperature leads to greater gas volume.
4. **Henry's law:** More gas pressure $\rightarrow$ greater gas solubility
5. **Diffusion:** Gases move from an area of higher pressure to an area of less pressure.

22.6.2 Physiology of Internal Respiration

Gas exchange direction depends on the pressure gradient between blood and tissues.

Oxygen Transport

The rate at which hemoglobin reversibly binds or releases oxygen is regulated by several factors: PO2, temperature, blood pH, PCO2, and BPG concentration.
1. An increase in partial pressure (pO_2) increases hemoglobin's affinity for oxygen.
2. Lower pH reduces hemoglobin's affinity for oxygen (Bohr effect).
3. An increase in carbon dioxide partial pressure (pCO_2) reduces hemoglobin affinity for oxygen.
4. Increased temperature reduces hemoglobin affinity for oxygen
5. Increased BPG reduces hemoglobin's affinity for oxygen.

All the above factors influence hemoglobin saturation by modifying hemoglobin's three-dimensional structure, thereby its affinity for oxygen. Increased temperature, PCO_2, H+, and BPG levels decrease hemoglobin's affinity for O2 and cause the oxygen-hemoglobin dissociation curve to shift to the right. This enhances oxygen unloading from the blood. All these conditions favor oxygen delivery to actively metabolizing tissues.

What Is Meant by "Shifting the Curve"?

"Shifting the curve" is a result of certain conditions, different from normal physiologic conditions, that
"move" the curve to the left or right

Transport of Carbon Dioxide in the Blood
1. Carbon dioxide is mostly transported as bicarbonate ions (HCO3). (70 percent)

$$CO_2 + H_2O = H2CO_3 \quad H^+ + HCO_3 \text{ (Carbonic anhydrase enzyme is}$$
required.)
Bicarbonate diffuses from RBC into the plasma and then to the lungs; this is counterbalanced by the movement of chloride ions into RBC. This is referred to as a

chloride shift.

2. Dissolved in plasma (7–10 percent)
3. Chemically bound to hemoglobin in the red blood cells

    CO_2 + hemoglobin = $HbCO_2$ (**carbaminohemoglobin**). The lower the PO_2 and the hemoglobin saturation with oxygen, the more carbon dioxide that can be carried in the blood (**Haldane effect**).

22.7 Control of Respiration

The basic controls of breathing involve the activities of neurons in the reticular formation of the medulla and pons. The medulla sets the rhythm.

The **medullary rhythmicity area** contains the inspiratory and expiratory areas. It sets the basic rhythm. This area communicates with the diaphragm via the phrenic and intercostals nerve.

The **pons** helps to switch between inspiration and expiration.

The **pneumotaxic area of the pons** limits inspiration and overrides the apneustic area.

The **apneustic area of the pons** limits expiration and stimulates inspiration. It works when the pneumotaxis area is inactive.

22.8 Factors Influencing the Rate and Depth of Breathing

Inflation reflex: Stretch receptors located in visceral pleurae and conducting passages in the lungs are stimulated when lungs are inflated. The receptors in turn send inhibitory impulses, via vagus nerves to the medullary respiratory centers, that end inspiration and allow expiration to occur.

Chemical factors: The chemoreceptors in the medulla oblongata monitor the H^+ ions. Peripheral receptors located in the carotid body and aortic body monitor the H^+ ions and O_2 levels.

22.9 Other Controlling Factors
1. Increased body temperature
2. Pain: acute or chronic
3. Emotional stimuli acting through the limbic system activate sympathetic centers in the hypothalamus.
4. Upper respiratory stimuli
5. Cortical influences:We can choose to hold our breath or take an extra deep breath.

Review Questions

1. **The portion of the pharynx that receives both air and food is the _____.**
2. **The openings to the auditory tubes are located in the _____.**
3. **The septa divide the lungs into _____.**
4. **The ease with which the lungs stretch in response to changes in pressure is termed _____.**
5. **When the oxygen concentration in the interstitial fluid declines, the condition is known as _____.**
6. **In the condition known as _____, a tissue's oxygen supply is cut off completely.**
7. **Quiet breathing is referred to as _____.**
8. **The volume of air moved in a quiet respiration is termed _____.**

9. The volume of air that can be forcefully expelled from the lungs following a normal exhalation is termed _____.

10. The volume of air that can be forcefully inhaled on top of a normal inspiration is termed _____.

11. The volume of air that remains in the lungs after a forced expiration is termed _____.

12. Damage to the septal cells of the lungs would result in _____.

13. When does "chloride shift" occur?

    _____

14. What is hyperpnea?

    _____

15. What is the function of the apneustic centers of the pons?

    _____

16. Boyle's law states:

    _____

17. What is pleurisy?

    _____

18. What is the function of the Hering-Breuer reflex?

    _____

19. Prolonged inspirations can result from stimulating the _____ center.

20. The most important chemical regulator of breathing is _____.

Chapter 23

The Digestive System

The organs of the digestive system can be separated into two main groups: (1) those of the alimentary canal and (2) accessory digestive organs.

23.1 The Alimentary Canal (Gastrointestinal (GI) Tract)

- The continuous, muscular digestive tube that winds through the body
- It digests food and absorbs the digested fragments through its lining into the blood.
- It forms direct links between organ.

Structures: mouth, oral cavity, pharynx, esophagus, stomach, duodenum, jejenum, ileum, cecum, ascending colon, transverse colon, descending colon, sigmoid colon, rectum, and anus

Accessory structures: teeth, tongue, salivary glands, liver, gall bladder, and pancreas

23.2 Digestive Processes

1. Ingestion: taking food into the digestive tract through the mouth
2. Propulsion: It includes swallowing (voluntary) and peristalsis (involuntary).
3. Mechanical digestion: physically prepares food for chemical digestion; includes chewing, mixing of food with saliva, churning, and segmentation.
4. Chemical digestion: series of catabolic steps in which complex food molecules are broken down to their chemical building blocks with the aid of enzymes
5. Absorption: passage of digested food, vitamins, minerals, and water from the lumen of the GI tract into the blood or lymph
6. Defecation: eliminates indigestible substances from the body via the anus in the form of feces

23.3 Histology of the Alimentary Canal

The walls of the alimentary canal have four basic layers, or tunics. From the lumen outward, these layers are the mucosa, submucosa, muscularis externa, and the serosa.

23.3.1 Mucosa

- Epithelial lining with enteroendocrine cells
  - The lamina propria (areola tissue) underlie the epithelium-isolated lymphoid tissue, part of MALT (the mucosa associated lymphoid tissue) important in the defense against pathogens.

  - External to the lamina propria is the muscularis mucosae, which increase surface area and produce local movements of the mucosa.

23.3.2 Submucosa

- Made up of loose connective tissue
- Contains the submucosal plexus, which control GI secretion via the autonomic nervous system

23.3.3 Muscularis

This layer is responsible for segmentation and peristalsis. It is made up of two types of muscles:

1. Skeletal muscle: This is responsible for swallowing
      It is found in the tongue, pharynx, and upper esophagus.
2. Smooth muscle: involuntary muscle made of two layers—inner circular layer and outer longitudinal layer. It produces propelling action when it shortens and a mixing action when it compresses.

The muscularis contains the myenteric plexus, which controls GI tract motility.

23.3.4 Serosa
The serosa is the outermost layer of the intraperitoneal organs (visceral peritoneum).
- It forms the mesentary, mesocolon, greater omentum, and lesser omentum.
- Retroperitoneal organs have both a serosa (on the inside facing the peritoneal cavity) and an adventitia (on the side of the dorsal body wall).

23.4 The Mouth-associated Organs
   a. Cheeks: lateral border
   b. Soft and hard palate: superior border
   c. Tongue: inferior
   d. Oral orifice: anterior opening
   e. Oropharynx: posterior opening continuous with the oral cavity
   f. Lined with stratified squamous epithelium

23.4.1 The Lips (Labia) and Cheeks
- The orbicularis oris muscle forms the bulk of the fleshy lips.
- The cheeks are formed by buccinators.
- The vestibules are the gums and teeth; within them is the oral cavity.
- The labial frenulum joins the internal aspect of each lip to the gum.

23.4.2 The Palate
- It forms the roof of the mouth.
- The hard palate lies anteriorly and the soft palate posteriorly.
- The uvula is the fingerlike projection from soft palate that closes off the oropharynx when we swallow.

23.4.3 The Tongue
- Occupies the floor of the mouth
- Made up of interlacing bundles of skeletal muscle fibers
- Contains both intrinsic and extrinsic skeletal muscle fibers

    Intrinsic: Not attached to bones
       Allow tongue to change shape (but not its position)—thicker, thinner, longer, or shorter

    Extrinsic: Extend to the tongue from their origins on bones of the skull or the soft palate. They alter the tongue's position, protrude it, retract it, and move it from side to side.

23.5 The Salivary Glands
A number of glands both inside and outside the oral cavity produce and secrete saliva.

Functions and Composition of
Saliva

1. Cleanses the mouth
2. Dissolves food chemicals so that they can be tasted
3. Moistens food so it can be made into bolus
4. Contains enzymes that begin the digestion of starch in the mouth
5. Saliva contains, water, amylase, lysozyme, ions, gases, area, and uric acid.

23.5.1 Types of Glands

1. Parotid: Lies anterior to the ear between the masseter muscle and the skin, The parotid (Stensen's) duct runs parallel to the zygomatic arch and opens into the vestibule next to the upper molar, Parotid glands secrete salivary amylase.
2. The parotid (Stensen's) duct runs parallel to the zygomatic arch and opens into the submandibular gland. It is near the base of the tongue at the angle of the mandible. Secretions pass through submandibular (Wharton's) duct. It secretes mucous and some amylase.
3. Sublingual gland: Located above the submandibular glands; contains the lesser sublingual (Rivinus's) duct

The salivary glands are controlled by both the sympathetic and the parasympathetic nervous system. The normal setting is the parasympathetic

23.6 The Pharynx: Food passes posteriorly into the oropharynx and then the laryngopharynx (common passageways for food, fluids, and air. It provides muscular contractions for swallowing.

23.7 The Esophagus: It serves as a passageway for food and muscular constrictions for swallowing. It passes through the mediastinum and then pierces the diaphragm at the esophageal hiatus and joins the stomach at the cardiac orifice.which is surrounded by gastroesophageal sphincter.

23.8 Deglutition (Swallowing)

It involves coordinated activity of the tongue, soft palate, esophagus, and several muscle groups. It has three major phases, the voluntary (buccal) phase, the pharyngeal phase, and the esophageal phase.

23.8.1 Voluntary Phase
- Occurs in the mouth
- The tip of tongue is placed against the hard palate, where it contracts and forces the food into oropharynx.
- Food stimulates tactile receptors in the pharynx and passes out of control.

23.8.2 Pharyngeal Phase
- Involuntary phase when food is swallowed
- Controlled by the swallowing center in the medulla and lower pons.
- The tongue blocks off the mouth.
- The soft palate rises to close off the nasopharynx.
- The larynx rises so that the epiglottis covers the opening into airways.

23.8.3 Esophageal Phase
- Food is pushed down as the pharyngeal muscles contract.
- The glossopharyngeal sphincter relaxes, allowing food to enter the stomach.

23.9 Esophagus

- **It passes through the esophageal hiatus of the diaphragm and ends at the stomach.**
- **It tansports food from the pharynx to the stomach.**
- **It has skeletal muscle in the superior part and smooth muscle in the inferior part.**
- **It has an upper esophageal sphincter and a lower esophageal sphincter which regulate the movement of materials into and out of the esophagus.**
- **The mucosal lining is moist stratified squamous epithelium.**
- **The submucosal layer produces mucus**

23.9 The Stomach

The stomach is the temporary "storage tank" where the chemical breakdown of protein begins and food is converted to chyme.

- The stomach has three muscle layers: oblique, circular, and longitudinal.

23.9.1 Regions
1. Cardiac sphincter: Food enters the stomach from the esophagus.
2. Fundus: dome-shaped part beneath the diaphragm
3. Antrum (pylorus)
4. Pyloric sphincter: controls stomach emptying

23.9.2 Microscopic Anatomy of the Stomach

Mucous Membrane
1. G cells make gastrin.
2. Goblet cells make mucous.
3. The gastric pit contains oxyntic glands made of parietal cells—HCl production.
4. Chief cells (zymogenic) produce pepsinogen and gastric lipase.
5. The stomach is highly vascular.
6. The inner surface is thrown into folds called rugae.
7. Contains enzymes that work best at pH 1-2

23.9.3 Functions of the Stomach
1. It mixes food. Peristaltic waves mix food with gastric juice.
2. Start digestion of protein, nucleic acid, and fats
3. It activates some enzymes such as pepsinogens.
4. It destroys bacteria.
5. It makes an intrinsic factor for B_{12} absorption, important in preventing pernicious anemia.
6. It absorbs alcohol, water, vitamin B_{12}, and aspirin.

23.10 Regulation of Gastric Secretion

Gastric secretion is controlled by both neural and hormonal mechanisms. The gastric mucosa produces about 3 L of gastric juice daily. Nervous control is provided by the vagus nerve and local enteric nerve reflexes. Hormonal secretions are under the control of gastrin, which stimulates the release of enzymes and HCl.

Phases of Secretions

23.10.1 Cephalic phase
- This phase is triggered by sight, smell, or thought of food.
- Input from these receptors are relayed to the hypothalamus.
- The hypothalamus stimulates the vagal nuclei of the medulla oblongata to increase secretion of gastrin, HCl, mucous, and pepsin.

23.10.2 Gastric Phase
- Presence of food in the stomach causes local neural and hormonal mechanisms to initiate the gastric phase.
- H ions are removed from the blood and HCO_3 ions are dumped into the blood (alkaline tide).
- Increased stomach secretion closes the cardiac sphincter and opens the distal sphincter.
- Gastric motility increases.

23.10.3 Intestinal Phase
- Food enters the duodenum, increasing release of enteric gastrin.

  - Increasing levels of H+, fats, and partially digested proteins trigger the inhibitory reflex (enterogastric reflex).

Enterogastrones
They trigger enterogastric reflex: secretin, cholecystokinin (CCK), vasoactive intestinal peptide (VIP), gastric inhibitory peptide (GIP)
Chyme: increases gastric section during the gastric phase and decreases gastric secretion during the intestinal phase (enterogastric reflex).

23.11 Regulation of Gastric Emptying
- The stomach usually empties completely within four hours after eating.
- Stimulated by nerve impulses caused by stomach distension and stomach gastrin
- It is limited by the rate of chyme processing in the small intestine.
- It is inhibited by the enterogastric reflex.
- A carbohydrate-rich meal moves through duodenum rapidly, but fats move more slowly.

23.11.1 The Small Intestine
The small intestine is a convoluted tube extending from the pyloric sphincter in the epigastric region to the
ileocecal valve.

Regions
1. Duodenum: mostly retroperitoneal; connected to pancreas via hepatopancreaticampulla
2. Jejenum
3. Ileum

Movements: Segmentation and peristalsis

Histology
1. Intestinal glands produce intestinal enzymes.
2. Duodenal glands produce alkaline mucous
3. Paneth cells produce lysozyme.
4. Microvilli: tiny projections of the plasma membrane of the absorptive cells of mucosa (brush border)
5. Lacteals: dense capillary and a wide lymphatic capillary in the villi.
6. Plica circularis: deep, permanent folds of the mucosa and submucosa.
7. Galt associated lymphatic tissue play a role in immunity
8. Small intestine absorb 80 percent of ingested water, electrolytes, vitamins, minerals, and carbohydrates (monosaccharides) by active/facilitated transport. Proteins are absorbed as dipeptides and amino acids. Lipids are absorbed as triglycerides, fatty acids micelles and omicrons.
9. The small intestine secretes sucroses, maltose, lactose, lipase, nuclease, and peptidases.
10. Small intestine is controlled by the parasympathetic nervous system and the presence of chime, CCK, and secretin.
11. Requires pancreatic enzymes and bile to complete digestion

23.11.2 Large Intestine
- Extends from ileocecal valve to anus
- Regions: Cecum—appendix
  Colon: Ascending → Transverse → Descending → Rectum → Anal canal

Histology

1. Villi absent
2. No permanent circular fold
3. The wall forms a series of pouches called haustra.
4. Three longitudinal bands of smooth muscle—taeniea coli
5. Epiploic appendages—numerous teardrop-shaped sacs of fat

Functions
1. Mechanical digestion: haustral churning and peristalsis; gastroileal and gastrocolic reflexes
2. Chemical digestion: bacteria ferment carbohydrate; protein/amino acid breakdown
3. Absorbs more water, vitamin B

23.11.3 Pancreas is located in the bend of the duodenum posterior to the stomach and liver.
Histology:
1. **Islet cells** perform endocrine functions.
2. Acini perform digestive functions. Its products drain into the pancreatic duct. The duct connects to the small intestine via the pancreatic duodenal duct (Ampulla of Vater).
3. Acini cells secrete pancreatic juice, which contains bicarbonate ions, amylase, nucleases, lipase, and proteases(trypsin, chymotrypsin, and carboxypeptidase).
4. Pancreatic secretion is controlled by the parasympathetic nervous system.
5. Chyme in the duodenum increases secretin secretion, bicarbonate ions.
6. Increased CCK stimulates pancreatic enzyme release.

23.11.4 The Liver
- Location: right hypochondrium and epigastric regions
- The Liver is divisible into left and right lobes, separated by the falciform ligament; associated with the right lobe are the quadrate and caudate lobes
- Each lobe contains hepatocytes that surround the sinusoids that feed into the central vein.

Functions of the Liver
1. The liver makes bile that acts as a detergent to emulsify fats. The release of bile is stimulated by the vagus nerve, CCK, and secretin.
2. The liver detoxifies and removes drugs and alcohol.
3. It stores glycogen, vitamins (A, D, E, and K), iron and other minerals, and cholesterol.
4. Activates vitamin D
5. Site of fetal RBC production
6. Carries out phagocytosis
7. Metabolizes absorbed food molecules
8. The liver has dual blood supply: the hepatic portal vein—direct input from small intestine.
9. Hepatic artery/vein—direct links to heart

23.11.5 The Gall Bladder
- The gall bladder lies under the quadrate lobe of the liver
- The gall bladder is involved in bile concentration, storage, and release.
- Contains smooth muscle that forces bile out after meals
- The hepatopancreatic ampulla (sphincter of Oddi) guards the entrance of bile into the duodenum.
- The major stimulus for gall bladder contraction is cholecystokinin (CCK).

Review Questions

1. The layer of loose connective tissue beneath the digestive epithelium is the

   _____.

2. Define the term peristalsis.

   _____

3. The portion of the tooth that receives blood vessels and nerves is the _____.
4. The space between the cheeks or lips and the teeth is called the _____.
5. The parietal cells secrete _____.
6. The chief cells secrete _____.
7. Enteroendocrine cells of the stomach secrete _____.
8. The prominent ridges in the lining of the stomach are called _____.
9. The pockets in the lining of the stomach that contain secretory cells are called

   _____.

10. What is the function of enzyme pepsin?

    _____

11. The structures that increase the surface area of the mucosa of the small intestine are
    called _____.
12. Peyer's patches are found in the _____.
13. An intestinal hormone that stimulates the gall bladder to release bile is

    _____.

14. An intestinal hormone that stimulates the release of insulin from the pancreatic islet
    cells is _____.
15. The function of enzyme enterokinase is to _____.
16. The fusion of the hepatic duct and the cystic duct forms the _____.
17. The basic functional unit of the liver is the _____.
18. The gastric phase of gastric secretion is triggered by the _____.
19. What are haustra?

    _____

20. What are taenia coli?

    _____

   Match the digestive juices and enzymes on the left with the descriptions on the
                                    right.
21. Amylase   enzymes that works on mlat sugars
22. bile   enzyme that works on cane sugar
23. lactase   enzyme found in the mouth
24. maltase   hydrochloric acid activates this enzyme
25. pepsin   enzyme that works on milk sugars
26. peptidases   enzyme that digests starch
27. sucrase   enzyme that converts maltose to glucose
28. trypsin   pancreatic enzyme that works on protein
                       major enzyme in stomach associated with protein
                       breakdown enzyme from lining of small intestine that
                       produces the end products of amino acids
                       Emulsifies fat
29. The hormone that stimulates the release of a bicarbonate substance from the pancreas
    is _____.
30. Chylomicrons are formed for the transport of _____.

Chapter 24
Metabolism

A nutrient is a substance in food that is used by the body to promote normal growth, maintenance, and repair. It is absorbed into the blood from GI tract.

24.1 Classes of Nutrients
1. Carbohydrates (monosaccharides): for energy production and modification of structures
2. Lipids: Lipids are absorbed as fatty acids, glycerol, and monoglycerides. They are used for energy storage, hormones, and structural components of cells and organs.
3. Proteins (amino acids): important structural components of the body like keratin of skin, collagen, and elastin. They regulate body functions as enzymes, hormones, and hemoglobin.
4. Minerals and vitamins: Most function as coenzymes.
5. Water: acts as solvent or suspension medium in chemical reactions. It participates in hydrolysis reactions. Water is also an important lubricant. It functions as a heat sink in the body.

24.2 Metabolism
Metabolism involves varieties of biochemical reactions of the body. It balances needs for energy and structure.

24.2.1 Divisions
1. Anabolism: Larger molecules are built from smaller ones, such as amino acids → proteins. May use dehydration synthesis reaction: Monosaccharide + Monosaccharide → Disaccharide; or Glycerol + 3 fatty acids → triglycerides
2. Catabolism: process that breaks down complex substances into simpler ones—e.g., hydrolysis of food. Cellular respiration is another example in which energy is released as ATP.

$$(Glucose)n-H_2O \rightarrow (glucose)n-1$$
$$(Dipeptide-H_2O \rightarrow aa1 + aa1$$

Role of ATP: transfers energy from catabolic reaction to anabolic reaction. ATP traps small part of the energy for the transfer.

24.2.2 Generation of ATP
1. **Substrate-level phosphorylation** is the process in which high-energy phosphate groups are transferred directly from phosphorylated compounds to ADP.
2. **Oxidative phosphorylation** is the process carried out by electron-transport proteins, forming part of cristae membranes in mitochondria.

24.3 Carbohydrate Metabolism
All carbohydrates we ingest, except for milk sugar (lactose) and small amounts of glycogen from meats, are derived from plants. They are absorbed in the form of monosaccharides and disaccharides.
1. They are absorbed by active transport and facilitated diffusion enhanced by insulin.
2. Glucose is trapped by the process of phophorylation as -Glucose 6-phosphate.
3. Glucose + ATP → glucose-6-phosphate + ADP
4. The trapping of glucose keeps intracellular glucose levels low, thereby maintaining a diffusion gradient for glucose entry.
5. Carbohydrates are used for ATP production, amino acid synthesis, glycogenesis, and lipogenesis.

24.3.1 Glucose Metabolism
   1. Oxidation (cellular respiration)
   2. Anaerobic: Sugar is converted into pyruvic acid (glycolysis) in the absence of oxygen.
   3. Aerobic: Involves the Krebs cycle and electron transport in the presence of oxygen

24.3.2 Stages of Glucose Metabolism

Glycolysis occurs in the cytoplasm of cells. It refers to the breakdown of the six-carbon molecule, glucose, into two three-carbon molecules of pyruvic acid or lactic acid if there is low oxygen Pyruvic acid is converted to acetyl coenzyme A and enters the Krebs cycle. Glycolysis produces 2NADH + 2H ions.+ 4 ATPs. 2 ATP molecules are used to prime the pump leaving a net gain of 2 ATP molecules.

The **Krebs (TCA) cycle** occurs in the aqueous mitochondrial matrix and is fueled by pyruvic acid.

1. **Decarboxylation:** Carbon is removed from pyruvic acid and released as carbon dioxide, which diffuses out of the cells into the blood to be expelled by the lungs.

2. **Oxidation:** Hydrogen is removed and picked up by NAD^+ - $NAD^+ + H \rightarrow NADH$.

3. The resulting acetic acid is combined with coenzyme A to produce **Acetyl coenzyme A (Acetyl CoA).**

4. **Acetyl CoA** now enters the Krebs cycle, where it is broken down completely by mitochondrial enzymes into acetic acid.

5. Coenzyme A shuttles the two-carbon acetic acid to the enzyme that condenses it with a four-carbon **oxaloacetic acid** to produce the **six-carbon citric acid** (the first substrate of the cycle).

6. The cycle moves along through its eight successive steps, atoms of citric acid being rearranged to produce different intermediate molecules called keto acids.

7. At the end of the cycle, acetic acid has been totally disposed of, and oxaloacetic acid, the substrate molecule, is regenerated.

End Products
   1. Production of 6NADH + 6 H ions (18 ATPs)
   2. Production of 2 $FADH_2$ (4ATPs)

24.4 Electron Transport Chain and Oxidative Phosphorylation

1. The electron transport chain oversees the final catabolic reaction that occurs in the inner mitochondrial membrane.

2. The hydrogens removed during oxidation of food fuels are finally combined with molecular oxygen.

3. Energy released during those reactions is harnessed to attach inorganic phosphate to ADP (this is oxidative phosphorylation): ADP + p1 $\rightarrow$ ATP

4. Components of the electron transport chain are proteins bound to metallic atoms (cofactors), which form part of the structure of the inner mitochondrial membrane (e.g., flavins and cytochromes). These are the electron carriers that are alternatively reduced and oxidized by picking up electrons and passing them to the next in the sequence.

5. Chemiosmosis: Ion channels and their attached coupling factors use the kinetic energy of passing hydrogen ions to generate ATP (creates H ion gradient).

6. Final outcome: Glucose + $O_2 \rightarrow CO_2$ + water + 36–38 ATPs

24.5 Glucose Anabolism

When more glucose is available than can be immediately oxidized, rising intracellular ATP concentrations eventually inhibit glucose catabolism and initiate processes that store glucose as glycogen or fat (anabolism).

1. **Glycogenesis:** storage of glucose in the liver and skeletal muscles when energy levels are high (glycogen formation). It is stimulated by insulin.
2. **Glycogenolysis:** conversion of glycogen to glucose when energy levels are low. It is stimulated by glucagon and epinephrine.
3. **Gluconeogenesis:** occurs when the liver is low in glycogen; converts proteins (amino acids), lactic acid, and fats (glycerol) into substances made in glycolysis. This process is stimulated by cortisol, thyroid hormone, epinephrine, glucagon, and growth hormone.

Note: Glucose anabolism helps to maintain blood sugar levels required for proper nervous functions.

25.6 Lipid Metabolism

- Ingested triglycerides are absorbed as fatty acids and monoglycerides.
- Small fatty acids are transported by direct diffusion.
- Larger fatty acids and monoglycerides are transported as micelles, chylomicrons, and lipoproteins
- Lipids are used for energy storage, energy use, steroid hormones synthesis, cell membrane formations, eicosanoids synthesis, and protein modification.
- Lipids are stored in adipose tissue in the subcutaneous tissue around the kidney, heart, genital areas, between muscles, and behind the eyes.

24.6.1 Lipid Catabolism

Lipids are required for cardiac muscle energy when glucose levels are low.

Lipolysis:

1. Triglycerides are broken down into glycerol and fatty acids.
2. Glycerol is converted to glyceraldehyde-2-phosphate → glucose if ATP is high (glycolysis).
3. Fatty acids are processed in the mitochondrial matrix.
4. Two carbon are removed from the fatty acid by ß- oxidation to make acetyl CoA. Acetyl CoA enters the Krebs cycle.
5. Excess acetyl CoA is processed to make ketone bodies: acetone, ecetoacetic acid, and ß-hyroxybutyric acid (ketogenesis).

24.6.2 Lipid Anabolism

Lipogenesis

Excess dietary intake is converted to triglycerides (lipoproteins, phospholipids, and cholesterol). The process occurs in the liver and adipose tissue. Lipogenesis is stimulated by insulin.

Note: Lipids are larger molecules than glucose; therefore their metabolism yields more pyruvates, and, therefore, more ATPs. Hence lipid is referred to as a "high energy" molecule. Fats have more calories per unit weight (9 calories/gm) than sugars (4 calories/gm).

24.7 Protein Metabolism

1. Proteins have a limited life span and must be broken down to their amino acids before they deteriorate.
2. Amino acids are absorbed into blood by active transport through the intestinal wall. Any excess is converted (through gluconeogenesis, lipogenesis, and ATP production). ,
3. Protein is used to make enzymes, hormones, antibodies, transport molecules and

contractile.elelments

24.7.1 Protein Catabolism

The body is able to recycle amino acids from catabolized proteins. The liver converts amino acids when energy sources are low.

Processes Involved

1. Deamination: Amino group (NH2) is removed to yield a pyruvate molecule.
2. Transamination: Amino acids transfer their amine group to α-ketoglutaric acid to form glutamic acid (amino acid ↔ keto acid).
3. Oxidative deamination: The amino acid group of the glutamic acid is converted into ammonia in the liver, and α-ketoglutaric acid is regenerated. The liberated ammonia molecules are combined with carbon to yield urea and water. Urea is eliminated in the urine.
4. Keto acid modification: Keto acids resulting from transamination are altered as necessary to produce metabolites that can enter the Krebs cycle (e.g., pyruvic acid, acetyl CoA, α-ketoglutaric acid, and oxaloacetic acid).
5. Amino acids enter cells by active transport as stimulated by insulin and growth hormone.

24.7.2 Protein anabolism involves the formation of peptide bonds. It is controlled by growth hormone, thyroid hormone, and insulin.

Note: Dietary protein is essential for growth, repair of body tissues, and pregnancy.

Nonessential amino acids are easily formed by siphoning keto acids from the Krebs cycle and transferring amine groups to them by transamination in the liver. Essential amino acids must be provided in the diet.

24.8 Absorptive and Postabsorptive States: Events and Controls

Metabolic controls act to equalize blood concentration between two nutritional states: fed state
(absorptive) and shortly after eating (postabsorptive state).

Absorptive State (proprandial) is the time soon after food is ingested. It requires about four hours depending upon the type and quantity of food ingested.

1. ATP is produced directly from glucose ingested.
2. **Carbohydrates** are absorbed as monosaccharides and delivered to the liver directly. Glucose is converted to glycogen and fat. Glucose that enters blood is metabolized for energy, and excess is stored in skeletal muscle cells as glycogen (glycogenesis).
3. **Triglycerides**: Most triglycerides enter the lymph in the form of chylomicrons. Adipose cells, skeletal muscle cells, and liver cells use triglycerides as an energy source when dietary carbohydrates are limited. Most enter adipose tissue, where they are reconverted to triglycerides and stored (lipogenesis).
4. **Amino acids**: They are delivered into the liver, where they are deaminated into ketoacids. The keto acids may flow into the Krebs cycle pathway to be used for ATP synthesis or may be converted to liver fat. Most amino acids remain in the blood to be used for protein synthesis.

Hormonal Regulation: Insulin stimulates cellular glucose uptake. Rising glucose levels (above 100 mg glucose/100 ml) in the blood act as a hormonal stimulus that prods the beta cells of the pancreatic islets to secrete more insulin.

Postabsorptive state (postprandial) is the fasting state after the ingested meal is absorbed. It maintains the blood glucose level by gluconeogenesis between 80–100 mg/100 ml.

Note: Most events of the postabsorptive state either make glucose available to the blood or to the organs that need it most (such as the nervous system).

Glucose sparing uses alternative energy molecules through the processes of glycogenolysis, lipolysis, and proteolysis. Glucose is reserved for the nervous system.

Hormonal control: Glucagon, cortisol, epinephrine/norepinephrine, growth hormone, and thyroid hormone

Note: An important stimulus for postabsorptive state is the damping of insulin release as the glucose level begin to drop and insulin antagonist, glucagon, release is stimulated (from the alpha cells of the pancreatic islets).

24.9 Regulation of Body Temperature

Homeothermic response keeps the body temperature at constant level.

1. Core (organs within the skull and thoracic/abdominal cavities) has the highest temperature.
2. Shell: Heat loss surface (skin and SubQ) has the lowest temperature

Hypothalamic Control

1. Preoptic area—heat-losing center: It is under parasympathetic control.
2. Heat-promoting center: It is under sympathetic control.

Mechanisms of Heat Production

1. Vasoconstriction: Blood is restricted to deep areas and largely bypasses the skin, thereby reducing heat loss.
2. Clinical thermogenesis: sympathetic stimulation increases cellular metabolism.
3. Shivering: involuntary muscle contraction
4. Thyroxine: when changing from hot to cold, the hypothalamus releases thyrotropin-releasing hormone. This hormone activates the anterior pituitary gland to release thyroid-stimulating hormone, which in turn induces the thyroid gland to secrete thyroid hormone (T3 and T4) into the blood, thereby increasing the metabolic rate.

Mechanisms of Heat Loss

1. vasodilation
2. perspiration
3. decreased metabolism

Abnormalities of Body Temperature

Fever is abnormally elevated body temperature caused by infection, tumor, heart attacks, and after surgery.

Mechanisms:

1. Phagocytes release interleukin-1.
2. Interleukin-1 causes release of prostaglandins.
3. Prostaglandins, in turn, reset the hypothalamic thermostat for a higher temperature.

Heat cramps occur when one loses water and sodium chloride in sweat.

Heat stroke occurs when the body can't eliminate heat, which can lead to brain death.

Hypothermia occurs when body temperature drops and cannot be raised. It may lead to death as a result of cardiac arrest.

Review Questions

1. **What is glycolysis?**

   _____

2. **The process of deamination produces _____.**
3. **_____ is the most important mechanism for the production of ATP.**
4. **The major anion in body fluids is _____.**

5.   The mineral that is necessary for the proper function of the enzyme carbonic anhydrase is _____.

Match the vitamins on the left with their functions on the right.

6.	Vitamin A	a constituent of coenzyme A
7.	Vitamin D	a constituent of the coenzyme NAD
8.	Vitamin E	a constituent of the coenzyme FAD and FMN
9.	Vitamin K	acts as an coenzyme in decarboxylation reactions
10.	Thimine	essential for the production of clotting factors
11.	riboflavin	prevents the destruction of vitamin A and fatty acids
12.	niacin	required for proper bone growth and calcium absorption
13.	Panthothenic acid	plays a role in maintaining epithelia and for visual pigment synthesis

14.   The process of synthesizing glucose from lipids, amino acids, or other carbohydrates is called _____.

15.   The process of glycogen formation is known as _____.

16.   _____ is the most important mechanism for the generation of ATP.

17.   The Krebs cycle takes place in the _____.

18.   List some events that occur during postabsorptive state.

_____

_____

_____

19.   The function of the citric acid cycle is to

_____.

20.   The liver forms glycogen during the _____ state.

Chapter 25
The Urinary System

Besides the urine-forming kidney, the urinary system includes the urinary bladder, the paired ureters, and the urethra.

25.1 Location and External Anatomy of the Kidney

1. The kidneys lie in a retroperitoneal position.
2. Usually lie just above the waist
3. The right kidney is lower than left kidney.
4. The adult kidney weighs about 150 g.
5. The hilus is directed toward the vertebral column.
6. The hilus is the entry site for blood vessels, lymphatic vessels, nerves, the ureter, and the renal sinus.
7. The kidney is surrounded by three tissue layer.

25.1.1 Internal Anatomy

A frontal section through a kidney reveals three distinct regions: the cortex, the medulla, and the pelvis.

1. The outer cortex contains renal columns.
2. The inner medulla contains cone-shaped tissue masses called renal pyramids.
3. The base of each pyramid faces towards the cortex; the apex (papilla) points internally.
4. The renal column is the inward extension of cortical tissue that separate the pyramids.
5. The lobe is the meullary pyramid together with surrounding capsule of cortical tissue.
6. The renal pelvis is located within the renal sinus and continuous with the ureter.
7. Major calyces are the branching extensions of the renal pelvis.
8. Minor calyces are the subdivisions of the major calyces that enclose the papillae of the pyramids

\

Nerve Supply

The nerve supply to the kidney and its ureter is provided by the renal plexus, a network of the autonomic nerve fibers and ganglia. It is largely supplied by sympathetic postganglionic fibers from the celiac plexus and the inferior splanchnic nerves and first lumbar splanchnic nerves. A renal nerve enters each kidney at the hilum. These are vasomotor nerve fibers that regulate blood flow by adjusting the diameter of the renal arterioles.

25.2 Nephrons

Each kidney consists of over 1 million tiny processing units called nephrons. There are also thousands of collecting ducts, each of which collects urine from several nephrons and convey it to the renal pelvis.

25.2.1 Structure of the Nephron

- Each nephron consists of a **glomerulus** associated with a renal tubule.
- The end of the renal tubule is called a glomerular (**Bowman's)** capsule, which surrounds the glomerulus.
- The **glomerular capsule** and the enclosed glomerulus are called the **renal corpuscle**.
- The glomerular endothelium is fenestrated, rendering it highly porous.
- The external parietal layer of the glomerular capsule is a simple squamous epithelium.

- The visceral layer in contact with the glomerulus consists of epithelial cells called **podocytes**.
- The extension of the podocytes makes contact with the basement membrane of the glomerulus.
- The clefts between the foot processes of the podocytes (filtration slits) allow the filtrate to pass into the capsular space.
- The remainder of the renal tubule has three parts: (1) proximal convoluted tubule (PCT), (2) loop of Henle, and (3) distal convoluted tubule (DCT).
- Each collecting duct receives urine from many nephrons.
- The collecting ducts fuse together to form large papillary ducts that empty into the minor calyces via the papillae of the pyramids.

25.3 Types of Nephrons

There are two types of nephrons. They are (1) **cortical nephron** located entirely in the cortex and (2) **juxtamedullary nephrons** that are located very close to the cortex-medulla junction. The ability of the kidneys to produce concentrated urine depend on the function of the juxtamedullary nephron.

Juxtaglomerular Apparatus

Each nephron has a region called the juxtaglomerular apparatus (JGA). This is where the initial portion of the distal tubule makes contact with the afferent arteriole (and sometimes the efferent). The walls of the arteriole consist of **juxtaglomerular (JG) cells** that secrete renin when the mechanoreceptors in them detect blood pressure changes. **Macula densa** cells adjacent to the juxtaglomerular cells are chemoreceptors that respond to changes in solute concentration in the filtrate.

25.4 The Filtration Membrane

The filtration membrane has three layers:
1. the fenestrated endothelium of the glomerular capillaries
2. the visceral membrane of the glomerular capsule, made up of podocytes.
3. the basement membrane composed of basal lamina of the other layers

Glomerular Filtration
- Forces fluids and dissolved contents through the membrane

Features Favoring Filtration
1. long glomerular capillaries
2. thin, porous filter
3. high glomerular blood hydrostatic pressure (GBHP)

Features Opposing Filtration
1. capsular hydrostatic pressure (CHP)
2. blood colloidal osmotic pressure (BCOP)

Net filtration pressure (NFP) = GBHP – (CHP + BCOP)

$$NFP = 55 \text{ mm Hg} - (30 \text{ mm Hg} + 15 \text{ mm Hg})$$
$$NFP = 10 \text{ mm Hg}$$

Filtration fraction is the percentage of plasma entering nephrons that becomes filtrate (16–20 percent)

25.4.3 Glomerular Filtration Rate (GFR)

GFR is the total amount of filtrate formed per minute by the kidneys. The normal GFR in both kidneys in adults is 125 mL/min or 180 L/day.

- GFR is affected by pressures that determine NFP.
- GBHP must be greater than 42 mm Hg.

Regulation of Glumerular Filtration

In humans, the GFR is held relatively constant by at least three important mechanisms that regulate renal flow: renal autoregulation, neural controls, and the renal-angiotensin system (hormonal).

1. Renal regulation operates by a negative feedback mechanism. It functions by directly regulating the diameter of the afferent and the efferent arterioles through a myogenic mechanism and a tubuloglomerular feedback mechanism. Low BP inhibits JGA, causing vasoconstriction.
2. Hormonal: The renin-angiotensin mechanism is triggered when juxtaglomerular (JG) cells of the juxtaglomerular apparatus (JGA) release renin in response to stimuli. Renin causes angiotensinogen to release angiotensin I. Angiotensin I is converted to angiotensin II by angiotensin-converting enzyme. Angiotensin II is a potent vasoconstrictor that stimulates aldosterone release.
3. Neural: The sympathetic nervous system is a vasoconstrictor.

Note: During stress, norepinephrine is released by sympathetic nerve fibers and epinephrine is released by the adrenal medulla. These hormones interact with alpha adrenergic receptors on vascular smooth muscle, causing strong vasoconstriction of the afferent arterioles, inhibiting filtrate formation.

25.5 Tubular Reabsorption

It is the reclamation process that returns materials from the filtrate to the blood. Ninety-nine percent of the filtrate is reclaimed and 1.5–2 L/day is lost as urine.

Mechanisms of Reabsorption

1. Reabsorption is by osmosis, active transport, pinocytosis, diffusion, and solvent drag.
2. Amounts reabsorbed depend on the body requirements and concentration gradient.

Proximal Convoluted Tubule (PCT)

- PCT is always permeable to water; 80 percent of water passing through is reabsorbed.
- Na, HCO3, K, Cl, and nutrients are reabsorbed by active transport.
- Urea and lipid-soluble solutes are reabsorbed by passive transport.

Descending Loop of Nephron

- Water is reabsorbed.
- Na ions and glucose are reabsorbed by active transport.

Ascending Loop of Nephron

- Chloride ions are reabsorbed by active transport.
- Na, K, and urea are reabsorbed by passive transport.

Distal Convoluted Tubule and Collecting Duct

- Water is reabsorbed when ADH is present.
- Na is reabsorbed by active transport when aldosterone is present.
- Anions and urea are reabsorbed by passive transport.

25.5.2 Tubular secretion adds materials from blood to the filtrate thereby helping to regulate pH and waste removal. It also helps to dispose of substances not already in the filtrate, such as drugs and excess potassium ions.

Proximal Convoluted Tubule (PCT)

- Cations H^+ and $NH4^+$ exchanged for Na and HCO3 ions
- Secretion of creatinine and antibiotics occur here.

Loop of Henle

- Secretion of urea occurs here.

Collecting duct

- K is exchanged for Na, regulated by aldosterone.
- plasma K concentration and distal convoluted tubule Na concentration
- H^+ is secreted against concentration gradient and buffered by NH3 and PO4.

Definition of Terms
Regulation of Urine Concentration and Volume

1. **Osmolality:** It is the number of solute particles dissolved in one liter (1000 g) of water, and it is **reflected in the solution's ability to cause osmosis**.
2. **Osmol**: 1 mole of nonionizing substance in 1 liter of water
3. **Milliosmol (mOsm)**: 0.001 osmol—normally used to describe solute concentration of body fluid. The kidneys tend to keep body fluids at a range of 300 mOsm.
4. **Countercurrent mechanism**: It means something is flowing in opposite directions through adjacent channels.
5. **Countercurrent multiplier:** flow of the filtrate through the long loop of Henle of juxtamedullary nephrons
6. **Countercurrent exchanger**: flow of blood through the limbs of adjacent vasa recta blood vessels

The Countercurrent Mechanism and the Medullary Osmotic Gradient

The mechanism of the countercurrent multiplication relies on the following requirements:

1. The structure of the cells of the loop of Henle varies in the length of the loop. Thus the loop is not equally permeable throughout its length.
2. The descending loop is freely permeable to water and impermeable to sodium and other solutes.
3. The ascending limb is impermeable to water, but can actively transport sodium and chloride ions out of the tubule.
4. There is the establishment of a concentration gradient in the interstitial fluid of the kidney, making interstitial fluid (ISF) of the cortex isotonic compared to the fluid in the tubule (300 mOsm). In the deep medulla, the ISF is hypertonic (1200 mOsm).
5. The structural arrangement of the loop of Henle and vasa recta is such that the tubes and vessels are in close proximity to each other. Both have a countercurrent arrangement.

Steps Involved in the Process

(i)   Na+ and Cl- leave ascending limb.
(ii)   Peritubular osmotic concentration increases.
(iii)   Water leaves descending limb, concentrating the filtrate.
(iv)   Na+ and Cl- exit is accelerated.

Benefits of Countercurrent Multiplication

(I) Efficient reabsorption of solutes and water

(ii) The resulting concentration gradient permits passive water reabsoption.

Dilution of urine occurs in the absence of ADH. Na, K, and Cl ions are actively reabsorbed in the ascending loop of nephron, distal convoluted tubule, and collecting duct.

Concentration of urine requires a high solute concentration in the interstitial fluid. ADH is required for excretion of concentrated urine.

Clinical Terminology

1. Calculi: insoluble deposits that form within the urinary tract from calcium and magnesium salts or uric acid
2. Aminoaciduria: amino acid loss in urine
3. Clearance test: a procedure used to estimate the glomerular filtration rate (GFR) by comparing plasma and renal concentration of a specific solute such as creatinine
4. Diuretics: drugs that promote fluid loss in urine
5. Glomerulonephritis: inflammation of the renal cortex
6. Hematuria: blood in the urine
7. Proteinuria: protein in urine
8. Renal failure: an inability of the kidneys to excrete wastes in sufficient quantities to maintain homeostasis

Review Questions

1. The outermost layer of kidney tissue is the _____.
2. Conical structures that are located in the renal medulla are called _____.
3. Describe the renal columns.

    _____

4. The expanded end of the ureter forms the _____.
5. The large branches of the renal pelvis are called _____.
6. Renal corpuscle is made up of _____ and _____.
7. The macula densa is part of a structure called the _____.
8. The cells of the macula densa and the juxtaglomerular cells form the _____.
9. The portion of the nephron closest to the renal corpuscle is the _____.
10. The primary function of the proximal convoluted tubule is

    _____.

11. The most selective pores in the filtration membrane are located in the

    _____.

12. The ability to form concentrated urine depends on the functions of the

    _____.

13. What is the role of "countercurrent multiplication" in the kidney?

    _____

    _____

    _____

14. What is the effect of increased level of ADH?

    _____

15 What is the effect of increased level of aldosterone on the kidney?

    _____

    _____

    _____

16. The _____ is responsible for the delivery of urine to the minor calyx.
17 Nephrons located close to the medulla with loops of Henle that extend deep into the renal pyramids are called _____.
18. The _____ is a capillary that surrounds the loop of Henle.
19. Explain the term trigone.

    _____

20. The _____ test is often used to determine the glomerular filtration.

Chapter 26

Fluid, Electrolytes, and Acid–Base Balance

26.1 Fluid Compartments of the Body

1. Intracellular fluid (ICF): About 67 percent of body fluid is in the cells .
2. Extracellular fluid (ECF): About 33 percent of body fluid is outside the cells. This includes:

   - Plasma (fluid portion of blood): 20 percent of ECF
   - Interstitial fluid: 80 percent of ECF (CSF, Lymph, and synovial fluid)

Body fluid is created and maintained by selectively permeable membranes.

26.1.1 Water

Water is the largest component of the body fluids.

- Percentage of water in adult males: 60
- Percentage of water in adult females: 50
- Percentage of water in infants: 73

26.2 Regulation of Fluid Levels

- Under normal conditions, fluid intake equals fluid output.
- Fluid is ingested as (preformed) water in liquids and foods.
- Fluid intake is regulated by dehydration, which stimulates the thirst center of hypothalamus to raise the blood osmotic pressure. This causes a drop in blood pressure, which causes the juxtaglomerular apparatus (JGA) in the kidney to release rennin.
- Metabolic water is derived from dehydration synthesis reaction (10 percent).

26.2.1 Water Output

1. Lungs: 28 percent of water vaporizes out of lungs as expired air.
2. Skin: insensible water loss (8 percent) 500 mL/day
3. Feces: 4%, about 200 mL/day
4. Kidneys: 60 percent as urine (1500 mL/day)

- Obligatory water loss: unavoidable water loss. It includes insensible water losses from lungs, skin, and feces.
- Sensible water loss: water loss in urine

26.2.2 Disorders of Water Balance

1. **Dehydration**: Water loss exceeds water intake over a period of time, and the body is in a negative fluid balance. It is caused by hemorrhage, vomiting, diarrhea, severe burns, profuse sweating, and diuretic abuse.
2. **Hypotonic hydration:** occurs when there is renal insufficiency or too much water is drunk very quickly, leading to cellular overhydration. This is referred to as water intoxication.
3. **Hyponatremia:** Sodium content in ECF is normal, but excess water is present (low ECF sodium level).
4. **Edema:** Atypical accumulation of fluid in interstitial space, leading to tissue swelling. It may be caused by increased blood pressure and capillary permeability, increased blood (capillary hydrostatic) pressure from incompetent venous valves, high blood volume, hypertension, and congestive heart failure.

26.2 Electrolytes Balance

Electrolytes include salts, acids, and bases, but the term electrolyte balance usually refers pH of the body. Electrolytes also carry electrical current.

26.2.1 Distribution

- Intracellular ions: K, Mg, PO_4, and protein anions
- Interstitial ions: Na, Cl, HCO_3, and protein anions
- Plasma ions: Na, Cl, HCO_3, and protein anions

26.2.2 Sodium is the most abundant extracellular cation.

- It determines the electrical gradient for impulse movement in nerves and muscles.
- It determines the osmotic pressure of ECF—water always follow sodium.

Regulation

- Aldosterone increases Na reabsorption.
- ADH: Low levels of ADH produce diluted urine, and high levels produce concentrated urine.
- ANP promotes water + Na excretion.
- Estrogen enhances NaCl reabsorption by the renal tubules, leading to water retention in pregnant women.
- Progesterone decreases sodium reabsorption by blocking the effect of aldosterone. It has a diuretic-like effect, promoting sodium and water loss.
- Glucocorticoids, such as cortisol and hydrocortisol, enhance tubular absorption of sodium, acting like aldosterone at high plasma levels.

26.2.3 Chloride (Cl) is the major extracellular anion.

- Can be found intracellularly
- Chloride is involved in balancing osmotic pressure.
- Chloride plays a role in the formation of Hcl for pH regulation.

Regulation: ADH—chloride ions follow Na ions.

Terms

Hypochloremia: low chloride level in blood as a result of overhydration, diuretic usage, hypokalemia, and emesis (vomiting)

Signs and symptoms: metabolic alkalosis due to bicarbonate retention, muscle spasms, depressed respiration, and coma

Hyperchloremia: high chloride level due to dehydration and hyperkalemia

Signs and symptoms: metabolic acidosis due to enhanced loss of bicarbonate, weakness, stupor, rapid deep breathing, and unconsciousness

26.2.4 Potassium (K) is the most abundant intracellular cation.

- Potassium determines the electrochemical gradient for impulses in muscles and nerves.
- Intracellular osmotic pressure
- pH regulation (K ions exchanged for H ions)

Regulation: Potassium level is regulated by aldosterone.

Terms

Hypokalemia: low blood potassium level caused by vomiting, diarrhea, renal disorders, excess sodium gain, and diuretic therapy

Signs and symptoms: Muscle cramps, fatigue, polyuria, flaccid paralysis, long Q-T interval, and flat T on ECG

Hyperkalemia: elevated blood K level

Signs and symptoms: weakness, irritability, and ventricular fibrillation

26.2.5 Calcium is the most abundant ion in the body. It is found extracellularly in bone and ECF.

- Calcium plays a vital role in muscle contraction.
- Blood coagulation
- Neurotransmitter release

- Excitability in nervous tissue and muscles

Regulation

- PTH: raises blood Ca level
- Calcitonin: lowers blood Ca level

Terms

- **Hypocalcemia:** low blood calcium level caused by high calcium loss, high phosphate level, and PTH disorder

Signs and symptoms: Tetany, convulsions, hyperactive reflexes, and paresthesia

- **Hypercalcemia:** elevated blood calcium

Signs and symptoms: lethargy, weakness, polyuria, slow reflex, emesis, altered mental state, coma, and death

26.2.6 Regulation of Phosphate (H_2PO_4, HPO_4, PO_4)

Locations

- Phosphate is found in intracellular and extracellular fluids.

Uses

- It provides compression strength for bone and teeth.
- It is a component of phospholipids, nucleic acids, sugar phosphates, and phosphoproteins.

Regulation

- PTH: increases blood phosphate
- Calcitonin: decreases blood phosphate

Terms

- **Hypophosphatemia:** depressed blood phosphate caused by urinary loss, alcoholism, and decreased absorption
- **Signs and symptoms:** loss of coordination, lethargy, seizures, chest and muscle pain, and confusion
- **Hyperphosphatemia:** elevated blood phosphate caused by excess intake and cell lysis
- **Signs and symptoms:** anorexia, tetany, tachycardia, nausea, and hyperactive reflexes

26.2.7 Regulation of Magnesium Balance

Locations

Magnesium is the second most abundant intracellular cation. Extracellular location includes bone and extracellular fluids.

Uses

- It activates the coenzymes needed for carbohydrate and protein metabolism.
- Neuromuscular activity and sodium pump

Regulation

- ADH
- PTH and AMP inhibit Mg^+ reabsorption in the PCT.

Terms

- **Hypomagnesium**: depressed blood level of magnesium caused by alcoholism, diarrhea, poor absorption, lactorrhea, and diabetes mellitus
- **Hypermagnesemia**: elevated blood magnesium level caused by renal failure, diabetic acidosis, dehydration, Addison's disease, and hypothermia
- **Signs and symptoms**: muscle weakness, flaccidity, emesis, and altered mental function

26.3 Acid–Base Balance

- pH depends on H+ ion concentration.
- H+ ions are produced by the metabolism of glucose, fatty acids, and amino acids.
- The normal pH of arterial blood is 7.4; that of venous blood and interstitial fluid is 7.35.

26.4 Maintenance Systems

26.4.1 Buffer

- A buffer is formed from a weak acid plus its salt.
- It temporarily binds H ions.
- Buffers work quickly.

26.4.2 Exhalation of Carbon Dioxide

- A change in the rate and depth of respiration changes pH.
- It works within a few minutes.

26.4.3 Kidney excretion

- Elimination of acids (except carbonic acid)
- Works in hours or days

26.5 Chemical Buffer System

26.5.1 Carbonic acid-bicarbonate

- Strong acid provides excess H ions; it is converted to weak acid (extracellular). $H + HCO_3 \rightarrow H_2CO_3 \rightarrow H_2O$
- Strong base (high pH) provides too few H ions; carbonic acid dissociates to bicarbonate ion. $H_2CO_3 \rightarrow HCO_3 + H$

26.5.2 Phosphate

- Low pH: $H + Na_2HPO_4 \rightarrow Na^+$ (Strong acid)(intracellular)
- High pH: $OH + NaH_2PO_4 \rightarrow H_2O$ (Strong base)

26.5.3 Protein (intra- and extracellular)

- Low pH: Amino groups bind H ions: $R\text{-}NH_2 + H^+ \rightarrow RNH_3^+$
- High pH: Carboxyl group dissociates H ion: $R\text{-}COOH \rightarrow RCOO^- + H^+$
- Histidine and cysteine have extra side groups that bind released H ions.

26.5.4 Hemoglobin

- $HbO_2 \rightarrow Hb + O_2$
- $Hb + H \rightarrow HbH$

26.6 Physiological Buffer Systems

- Exhalation of CO2: $\uparrow CO_2 \rightarrow \downarrow pH \downarrow CO_2 \rightarrow \uparrow pH$
- Respiratory centers of medulla react to chemoceptors.

  Increase ventilation→Blow off CO_2→ Reduce H ions →raise pH

  Decrease ventilation-→Retain CO_2→Retain H ions-→ lower pH

26.6.1 Kidney Excretion

- Kidneys eliminate lactic acid, uric acid, sulfuric acid, keto acids, and phosphoric acid. They reabsorb and generate HCO_3.

26.6.2 Acid–Base Imbalance

Compensation: The body tries to return pH to its normal range.

- Respiration compensates for metabolism.
- Metabolism compensate for respiration.

Effects

a. Acidosis (pH less than 7.35) leads to depressed CNS activity, coma, and death.
b. Alkalosis (pH greater than 7.45) leads to hypersensitive nervous system, nervousness, and muscle spasms.

pH disorders

- Respiratory disorder involves CO_2.
- Metabolic disorder involves HCO3.

Review Questions

1. **If the blood is alkaline, the body will compensate in a way to**
   _____
   _____
   _____.

2. **The role of the kidney in the maintenance of acid–base balance is to**
   _____.

3. **A buffer system is often composed of**
   _____.

4. **A decrease in the amount of carbon dioxide in the blood will _____ of the blood.**

5. **The most important buffer system in the plasma and intracellular fluid is the**
   _____.

6. **The hormone released by cardiac muscle as a result of abnormal stretching of the heart wall is _____.**

7. **If hydrogen ions are added to a buffer system, they react with the buffer system to form _____**
   _____
   _____.

8. **List some physiological changes that occur if a person is suffering from acidosis.**
   _____
   _____
   _____
   _____
   _____
   _____

9. A loss of alkaline reserve without significant change in blood pH is called
   _____.
10. Hyperventilation will result in _____.
11. List some events that would occur if the blood is acidic.

   _____

   _____

   _____

12 Some symptoms of alkalosis include: _____, _____, _____ and
   _____

13. Holding your breath for an extended period of time will result in
   _____.

14. Accumulation of dissolved carbon dioxide is known as _____.
15. As blood osmotic pressure increases, effective filtration pressure at the
   capillaries _____.
16. Prolonged hyperventilation is accompanied by _____.
17. A person with diabetic acidosis would have high/low level of respiration.
   (choose one)
18. Prolonged diarrhea can produce an acid–base imbalance classified as
   _____.

19. The sodium level of the blood is controlled primarily by the
   _____ hormone.
20. Write the formula for the Starling's law of capillaries.

   _____

21. As hydrogen ions (H+) increase, blood pH _____.
Match the electrolyte found in the left hand column with its function and importance
   listed
right hand column.

22.	Calcium	the major extracellular ion important in establishing the low pH ofgastric juice
23.	Phosphate	involved in blood coagulation and neurotransmitter release
24.	Potassium	a major intracellular ion that activates enzymes involved in carbohydrate metabolism
25.	Sodium	component of nucleic acids, some lipids, and high-energy storage compounds
26.	Chloride	intracellular ion important in establishing the polarity of the cell membrane
27.	Magnesium	extracellular ion important in establishing the polarity of the cell

Chapter 27
The Reproductive System

27.1 Anatomy of the Female Reproductive System

The functions of the female reproductive system include sex hormones production, functional gametes production, and protection and support of the developing embryo.

External Structures

Vulva (pudendum): mons pubis, labia majora, labia minora, clitoris, vestibule, external urethral orifice, paraurethral glands, greater vestibular glands, lesser vestibular glands, perineum, and mammary glands

Internal Structures

Ovaries, uterine tube (fallopian tube), uterus, and vagina

The Ovaries

- The ovary is a female gonad.
- It moves into the pelvis during the third month of development.
- Three ligaments hold it along the pelvic brim: the broad, ovarian, and suspensory ligaments

Histology of Ovary

- The **visceral peritoneum**, or **germinal epithelium** covers the surface of the ovary.
- **Tunica albuginea:** a dense connective tissue-capsule
- The **stroma** consists of a superficial cortex and a deeper medulla.
- **Gametes** are produced in the medulla.
- **Ovarian follicle:** immature ova
- **Vesicular (Graafian) follicle:** stimulated by estrogen to produce ovum
- The **corpus luteum** produces relaxin, inhibins, progesterone, and estrogens.

The **Uterine (Fallopian) tube** is the structure that transports ova. It consists of the following parts.
1. The fimbriae are the structure that collects ova.
2. The infundibulum is the open, funnel-shaped structure bearing fimbriae.
3. The ampula is the usual site of conception.
4. The isthmus is the largest part that joins the uterus.

The uterine tube has three layers:
1. Mucosa—lined with cilia (columnar epithelium)
2. Muscularis
3. Serosa

The **Uterus** has three main functions:
1. It provides support for developing offspring.
2. It expels the developed fetus during labor.
3. It is the site of menstruation.

The uterus is located between the bladder and the rectum and may be oriented in two positions:

1. **Anteversion:** Uterus bends anteriorly near the base.
2. **Retroversion:** Uterus bends backward toward the sacrum.

Layers of the Uterus
1. **Endometrium** is the vascular lining (functional layer).
2. **Myometrium** is the muscular layer.
3. **Perimetrium** is the visceral peritoneum.

Regions of the Uterus
1. The fundus is the rounded portion of the uterus.
2. The body is the largest portion of the uterus.
3. The cervix is the inferior portion of the uterus that extends from isthmus to the vagina.
4. The isthmus is the constriction on the uterus.
5. The cervical os is the external orifice of the uterus.
6. The cervical canal is the constricted passageway that opens into the uterine cavity at the internal os.

Ligaments
1. Broad
2. Cardinal
3. Round
4. Uterosacral

Blood supply
- Uterine artery arising from the branch of iliac artery
- Uterine vein—drains the uterus

The **Vagina** is the pliable and stretchable female copulatory organ that lies between the urinary bladder and the rectum.

Parts of the Vagina
1. The fornix is the shallow recess surrounding the cervical protrusion.
2. The hymen is the incomplete partition in the vaginal orifice.

Histology of the Vagina
- Lined with stratified squamous epithelium (nonkeratinized) overlying connective tissue (they form the rugae).
- Dendritic cells act as antigen-presenting cells—provide route for HIV transmission

The **Mammary Gland (Breast)** is a modified glandular tissue that contains milk-producing glands called the **alveoli.**

Route of Milk
1. **Secondary tubules**
2. **Mammary ducts**
3. **Lactiferous ducts**
4. **Nipple (papilla)**
5. **Areola (pigmented area around nipple) (Sebaceous glands make it bumpy)**
- Development of the breast is under hormonal control (**estrogen and progesterone**).
- Lactation: **Secretion is controlled by prolactin, estrogen, and progesterone. Ejection is stimulated by oxytocin.**

27.5 Female Reproductive Cycle

This occurs in non-pregnant females under hormonal control. The cycle length varies, the average length being about one month.

Primary events

- Maturation of ovum: ovarian cycle
- Preparation of endometrium: menstrual cycle

27.5.1 Hormonal Regulation

1. Hypothalamus (GnRH) → Adenohypophysis secretes: FSH—stimulates initial ovarian follicle development and estrogen secretion
2. LH—Stimulates ovulation and increased production of estrogen, inhibin, relaxin, and progesterone
3. Estrogens: for development of female reproductive system—e.g., endometrium, breast, secondary sex traits. Estrogen also functions in fluid/electrolyte balance and protein anabolism.
4. Progesterone is a synergist of estrogen.
5. Inhibin inhibits FSH, GnRH, and LH.
6. Relaxin softens pubic symphysis and assists in uterine cervix dilation.

27.6 Phases of the Female Reproductive Cycle

1. **Menstrual phase usually lasts from day 1 to 5.**
   - This is when the uterus loses blood, mucous, and epithelial cells. This is caused by a rapid drop in estrogen and progesterone levels.
   - Ovary: Primary follicles begin development. Zona pellucida forms around secondary oocyte. Follicular fluid forces oocyte to the edge of follicle. Ovarian development is under the control of FSH and GnRH.

2. **Preovovulatory phase (days 6 to 13)** comprised of two phases:
   - Proliferative phase: Estrogen stimulates endometrium build up.
   - Follicular phase: Secondary follicle matures.

3. **Ovulation: Day 14**
   - Vesicular ovarian follicle ruptures.
   - Occurs due to positive feedback of estrogen on LH and GnRH levels

Signs of ovulation

- LH surge precedes follicle rupture.
- Basal temperature rises.
- Cervical mucous thins.
- The cervix softens.
- Pain in the lower abdomen (Mittleschmerz)

4. **Postovulationary phase (days 15 to 28) consist of the following events:**
   - Luteal phase: Corpus luteum develops and begins to secrete estrogen and progesterone.
   - If there is no zygote, GnRH and LH are inhibited, estrogen and progesterone levels fall, and menses begins.

27.7 Anatomy of the Male Reproductive System

1. External structures: penis, scrotum, testes, epididymis, and part of ductus deferens
2. Internal structures: seminal vesicles, prostate gland, bulbourethral glands (accessory glands), and abdominal portion of ductus deferens

Scrotum
- The scrotum is a sac of skin and superficial fascia that hangs outside the abdominopelvic cavity.
- The testes are suspended inside the testes within the scrotal cavities.
- Contains two sets of associated muscles:
    1. **Dartos** wrinkles the skin.
    2. **Cremaster** (from internal oblique muscles) elevates testes.
- The superficial location of the scrotum provides a temperature 3° C lower than core body temperature, necessary for sperm production.

Testes
- Testes are oval, paired gonads approximately 4 cm long and 2.5 cm in diameter.
- They develop on the posterior abdominal wall and descend to the scrotum during the last two months of development

Histology of the Testes
1. **Tunica vaginalis:** two-layered tunic derived from the peritoneum
2. **Tunica albuginea:** fibrous capsule of the testis divided into lobules by septa
3. **Semineferous tubules** is the actual sperm factory where spermatogenesis occur.
4. **Tubulus rectus** is the straight tubule that convey the sperm into the rete testis.
5. **Sustentacular (nurse)** cells form the blood-testis barrier.
6. **Interstitial cells (leydig):** These are the cells that produce androgens **(testosterone)** and **inhibin.**

Terminology
Cryptorchidism: nondescent of the testes

27.7.2 The Penis
1. The penis is a copulatory organ designed to deliver sperm into the female reproductive tract.
2. The male perineum is the diamond-shaped region located between the pubic symphysis, coccyx, and ischial tuberosities.
3. The root is the fixed portion that attaches the penis to the body wall.
4. The glans is the exposed distal end that surrounds the external urethal orifice.
5. The prepuce surrounds the tip of penis. It attaches to the neck of the penis; it is the part that is removed during circumcision.
6. The body contains paired dorsolateral corpora cavernosa erectile tissues and one midventral corpora spongiosum (erectile body).

27.8 The Male Duct System
The sperm travel from the testes to the outside of the body through a system of ducts. In order
(proximal to distal), they are the **epididymis**, the **ductus deferens**, and the **urethra.**

Ducts Associated with Testes
1. Straight tubules of the testes (**tubulus rectus):** convey sperm into rete testis.

2. The **rete testis** is the sperm storage site
3. The **epididymis** is lined with pseudostratified columnar epithelium with stereocilia. The sperms mature and are stored here.
4. The **ductus deferens** runs as part of the spermatic cord from the epididymis through the iguinal canal into the pelvic cavity. The expanded terminus of the ductus deferens serves as a sperm storage site. It is lined with pseudostratified columnar epithelium and contains thick layers of smooth muscle that squeeze the sperm forward.
5. The **ejaculatory Duct** is located at the junction of the seminal vesicle duct and the ductus deferens. It plays a role in moving the sperm forward.
6. The **urethra** is the terminal portion of the male duct system. It conveys both urine and semen (at different times). The regions of the urethra include:
    *   The **prostatic urethra** is the portion surrounded by the prostate gland.
    *   The **membranous urethra** is the urogenital diaphragm.
    *   The **spongy (penile) urethra** runs through the penis and opens to the outside at the external orifice.

27.9 Accessory Glands

The accessory glands include the paired seminal vesicles, bulbourethral glands, and the single prostate gland. These glands produce the bulk of the semen.

1. The **seminal vesicles** are located on the posterior wall of the bladder. Their secretions account for about 69 percent of the volume of the semen. The fluid produced is yellowish, viscous, and alkaline. It contains fructose sugar, ascorbic acid, a coagulating enzyme, and prostaglandins.
2. The **prostate gland** encircles the part of the urethra just inferior to the bladder. It produces milky and acidic fluid that accounts for one-third of the semen volume. The fluid contains citrate and several enzymes, such as fibrolysin, hyaluronidase, acid phosphatase, and prostate-specific antigen (PSA).
3. The **bulbourethral gland (Cowper's gland)** is situated inferior to the prostate. It produces thick, clear mucus prior to ejaculation. The mucus lubricates the penis.

Review Questions

1. What is the function of inhibin?
    _____.

2. The hairless area of skin between the bottom of the labia and the anus is called the _____.

3. What is the function of the interstitial cells (Leydig)?
    _____

4. The developing sperm cells that go through the first meitotic division are _____.

5. A normal mature human spermatozoa contain _____ chromosomes.

6. The primary spermatocyte will eventually form _____ spermatozoa

7. The function of the sustentacular (Sertoli) cells in the testes is _____.

8. Prior to ejaculation sperm are stored in the _____.

9. The main control center of male reproductive processes is the _____.

10. The perineum is the area between the _____ and _____.

11. The membrane covering the vaginal opening is called the _____.
12. Corpus luteum produces which hormone? _____
13. The inner lips surrounding the vagina opening are the _____.
14. The innermost layer of the uterus is called the _____.
15. The three openings found in the vestibule are the _____, _____ and _____.
16. The finger-like structures at the open end of the fallopian tubes are the _____.
17. The structure that is homologous to the scrotum of the male in female is the _____.
18. What is the function of LH?

    _____
19. Name the two hormones produced by the ovaries.

    _____
20. Sperm develop from stem cells called _____.
21. The erectile tissue that is located on the anterior surface of the penis is the _____.
22. The surge of LH during the middle of the ovarian cycle triggers _____.
23. Menstruation is triggered a by the drop in the levels of _____ and _____.
24. The onset of menstruation at puberty is called _____.

Chapter 28
Pregnancy and Human Development

Definition of Terms

1. Pregnancy: events that occur from the time of fertilization until the infant is born
2. Conceptus: the pregnant woman's developing offspring
3. Gestation period: the time during which development occurs
4. Embryo: developmental stage extending from gastrulation to the end of the eighth week
5. Fetus: developmental stage extending from the ninth week of development to birth

28.1 Fertilization

- Fertilization is the process by which a sperm cell attaches to a secondary oocyte.
- Sexual intercourse must occur no more than three days before ovulation and no later than 24 hours after for fertilization to occur.
- Fertilization occurs approximately one-third of the way down the length of the uterine tube **(ampulla)**.
- The sperm penetrates the secondary oocyte cytoplasm and joins the oocyte proneuclues to form a zygote.

Steps

1. Sperm transport and **capacitation**: Only a few hundred sperm out of millions reach the uterine tube.
2. **Capacitation** of the sperms occur when the membrane becomes fragile enough for the hydrolytic enzymes in their acrosomes to be released.
3. **Acrosomal reaction and sperm penetration**: Hyaluronidase, acrosin, and protease enzymes are released to break down the **corona radiata** and **zona pellucida**.
4. A single sperm makes contact with the oocyte's membrane receptors, and its nucleus is pulled into the oocyte's cytoplasm
5. **Blocks to polyspermy: Depolarization** of the oocyte's membrane occurs after contact, preventing other sperm from fusing with the oocyte's membrane.
6. Ionic calcium (Ca^{2+}) is released by the oocyte's endoplasmic reticulum into the cytoplasm to activate cell division.
7. Completion of meiosis II: The sperm loses its tail and midpiece and migrates to the center of the oocyte.
8. The secondary oocyte completes meiosis II to form the ovum nucleus and ejects the second polar body.
9. The ovum and sperm nuclei swell to become male and female pronuclei, forming mitotic spindle as they approach each other.
10. The pronuclei membranes rupture, releasing the chromosomes.
11. Fertilization occurs as the paternal and maternal chromosomes combine and produce a diploid zygote.

28.2 Pre-embryonic Development

This process begins with fertilization and continues as the pre-embryo travels through the uterine tube and finally into the uterine cavity. Events involved are **cleavage** and **implantation**.

1. **Cleavage** is the period of rapid mitotic cell divisions of the zygote following fertilization. Daughter cells become smaller and smaller (high surface-to-volume ratio).

    - **Blastomeres** (two identical cells) are formed 36 hours after fertilization.
    - **Morula**: Sixteen or more cells are formed 72 hours after fertilization.
    - **Blastocyst:** fluid-filled hollow sphere composed of a single layer of large, flattened cells (**trophoblast** cells) and inner cell mass at one side.

Note: Trophoblast cells take part in placenta formation.

Terms

Totipotent cells are cells in the early stages of development (days 1-4) that have the potential to give rise to any type of tissue.

Differentiation: change of a cell in structure and function (specialization)

Pluripotent: After differentiation, any cell has the ability to develop into different tissues.

The inner cell mass becomes the embryonic disc, which forms the embryo proper.

2. **Implantation** begins six to seven days after ovulation. Receptivity of the endometrium for implantation depends upon the levels of the ovarian hormones estrogen and progesterone.

    - The trophoblast cells overlying the inner cell mass adhere to the endometrium.
    - The trophoblast proliferates and forms two distinct layers: **cytotrophoblast** (inner layer) and **syncytitrophoblast** (outer layer).
    - The outer layer rapidly digests the endometrium until the entire blastocyst is sealed off from the uterine cavity.
    - The viability of the corpus luteum is maintained by human chorionic gonadotropin hormone (hCG) secreted by the trophoblast cells of the blastocyst.
    - hCG bypasses the pituitary-ovarian controls and prompts the corpus luteum to continue to secrete progesterone and estrogen, maintain the pregnancy until the end of the second month.

Placentation refers to the formation of the placenta. The proliferating trophoblast gives rise to an extra-embryonic mesoderm in the inner surface, which becomes the chorion. The chorion develops fingerlike chorionic villi. The mesodermal cores of the villi become highly vascularized and extend to the embryo to form the arteries and veins. The continued erosion of the endometrium produces lacunae or inter villus spaces in the stratum functionalis of the endometrium. The villi come to lie in the spaces totally immersed in maternal blood.

Terms

- **Decidua basalis**: the part of the endometrium that lies between the chorionic villi and the stratum basalis
- **Placenta:** the **cho**rionic villi and the decidua basalis

Germ Layer Formation

The two-layered embryonic disc transforms into a three-layered embryo containing the primary germ layers—ectoderm, mesoderm, and endoderm—formed by the process of

gastrulation. The three primary germ layers are the primitive tissues from which all body organs will derive.

Derivatives of the Primary Germ Layers

Ectoderm
All nervous tissue
Epidermis of skin and its derivatives
Cornea and lens of eye
Epithelium of oral and nasal cavities

Mesoderm
Skeletal, smooth, and cardiac muscle
Cartilage, bone, and connective tissues
Blood, bone marrow, and lymphoid tissues
Ureters, kidneys, gonads, and reproductive ducts

Endoderm
Epithelium of digestive tract
Liver and pancreas
Epithelium of respiratory tract, auditory tube, and tonsils
Thyroid, parathyroid, and thymus glands

Fetal Growth
- After the eighth week, the developing human is called a fetus, and the period of time from then until the birth is called the fetal period.
- During the fetal period, organs established by the primary germ layers grow rapidly.

Gestation
- Gestation is the time a zygote, embryo, or fetus is carried in the female reproductive tract.
- Human gestation lasts about 266 days, counted from the estimated day of fertilization (280 days from the first day of the last menstrual period).

28.3 Hormones of Pregnancy
1. Pregnancy is maintained by **human chorionic gonadotropin (hCG)** by preventing degeneration of the corpus luteum. The levels of **estrogen** and **progesterone** are kept high, thus preventing menses.
2. hGC mimics LH in order to stimulate the corpus luteum to continue estrogen and progesterone production. These hormones maintain the lining of the uterus to ensure continued attachment of the embryo.
3. Human chorionic somatomammotropin (hCS), also known as human placental lactogen or hPL, is also produced by the chorion to control breast development for lactation, protein metabolism, and catabolism of glucose (glucose-sparing effect) and fatty acids.
4. Relaxin is produced by the placenta and ovaries for symphysis pubis relaxation and cervix dilation.
5. Inhibin, produced by the ovaries, testes, and placenta, inhibits secretion of FSH and regulation of hGH secretion.

Note: Pregnancy tests detect minute amounts of hCG. Pregnancy kits contain **antibodies**

to hCG and other chemicals that produce a color change if there is a reaction between hCG in the urine and the hCG antibody in the test kit.

28.4 Anatomical and Physiological Changes during Gestation

1. **Uterus:** Uterine growth in the first trimester occurs in response **to** hormonal stimulus of high levels of estrogen and progesterone. Enlargement results from (1) increased vascularity and dilatation of blood vessels, (2) hyperplasia (production of new muscle fibers) and hypertrophy (enlargement of preexisting muscle fibers), and (3) development of the decidua. During the early weeks of pregnancy, the uterus, cervix (**Godell's sign),** and isthmus soften (**Hegar's sign),** and the cervix takes on a bluish color (**Chadwicks's sign**). The fundus becomes easily flexed on the cervix (**McDonald's sign**).

2. **Weight gain,** increased protein, fat, and mineral storage. Maternal requirements for nutrient and vitamins increase by about 10–30 percent.

3. **Breast enlargement** due to the effects of the combination of human placental lactogen and placental prolactin from the placenta. The fullness, heightened sensitivity, tingling, and heaviness of the breast are preemptive signs of pregnancy. Hypertrophy of the sebaceous glands (**Montgomery's tubercles**) may be seen around the nipples.

4. **Cardiovascular modifications** include an increase in stroke volume by approximately 30 percent and a rise in cardiac output by approximately 30–50 percent. There is increased plasma volume which results in hemodilution (may result in physiological anemia). The increased blood volume is a protective mechanism. It compensates for (1) the hypertrophied vascular system of the enlarged uterus, (2) adequate hydration of fetal and maternal tissues when the woman assumes an erect or supine position, and (3) fluid reserve for fluid and blood loss during the birth and puerperium. There is an increase in white blood cells (primarily the granulaocytes) and clotting factors. Lymphocyte count stays the same throughout pregnancy.

5. **Marked alterations in pulmonary function** occur due to functional and structural adaptations that accompany pregnancy. The acceleration in metabolic rate and the need to add to the tissue mass in the uterus and breasts increases blood PO_2 and elevates the Pco_2. The combination stimulates the production of renin and erythropoietin, leading to increased maternal volume. Dyspnea also occurs as result of pressure on the diaphragm as the fetus increases in size. Edema and hyperemia may occur within the nose, pharynx, larynx, trachea, and bronchi due to elevated levels of estrogen that causes the capillaries to become engorged. The increased vascularity also swells the tympanic membranes and Eustachian tubes, giving rise to symptoms of impaired hearing, earaches, or a sense of fullness in the ears. There is marked increase in tidal volume (30–40 percent) and decreased expiratory reserve volume (by up to 40 percent).

6. Increased appetite during the second trimester and decreased motility that can result in constipation and delayed gastric emptying. Nausea, vomiting, and heartburn also occur. The basal metabolism rate (BMR) rises by the fourth month of gestation due to increased oxygen demand by the fetus, uterus, placenta, and increased maternal cardiac work.

7. Frequency and urgency of urination due to pressure on the urinary bladder.

Glomerular filtration rate rises up to 40 percent. There is a change in the maternal acid–base balance due to the fall in PCO2 and increased HCO3. These changes raise the blood pH (respiratory alkalosis), which may be compensated for by mild metabolic acidosis.

8. **Increased pigmentation of the skin, striae gravidarum** (stretch marks) over the abdomen occurs as the uterus enlarges, and hair loss. **Linea nigra** and face mask (**cloasma**) are other observed changes in the skin of a pregnant woman. **Angiomas** (vascular spiders) usually occur on the neck, thorax, face, and arms due to elevated levels of circulating estrogen. **Epulis** (gingival granuloma gravidarum), a lesion in the teeth, may occur around the third month of pregnancy. Nail growth accelerates during pregnancy. Pregnancy is also associated with oily skin and acne vulgaris.

9. **Edema and increased vascularity of the vulva** and increased pliability and vascularity of the vagina.

10. The gradually changing body and increasing weight of the pregnant woman confer pronounced effects on the **musculoskeletal system,** such as change in posture and walking ("the proud walk of pregnancy").

11. **Pregnancy-induced hypertension (PIH)** occurs among 15 to –15 percent of pregnant women in the United States. This is due to pre-eclampsia due to impaired renal function. It is called eclampsia when it is associated with convulsions and coma. Gestational hypertension is due to vasospasm caused by the secretion of angiotensin II, which causes the release of prostaglandin and throboxane (both powerful vasoconstrictors).

28.5 Labor

Labor is the process of moving the fetus, placenta, and membranes out of the uterus and through the birth canal.

- Decrease in the levels of progesterone and elevated levels of estrogen, prostaglandins, oxytocin
    (OT), and relaxin initiate the process of labor.
- Hormones produced by the normal fetal hypothalamus, pituitary gland, and adrenal cortex may also contribute to the onset of labor.
- True labor begins when uterine contractions occur at regular intervals, usually producing pain.
- Discharge of blood-containing mucus ("show") from the cervical canal
- Dilation of the cervix

Stages of Labor

The course of normal labor consists of the following:

1. Regular progression of uterine contractions
2. Effacement and progressive dilatation of the cervix
3. Progress in descent of the presenting part

First Stage of labor

The first stage of labor usually lasts from the onset of regular uterine contractions to full dilatation of the cervix. The first stage has been divided into three phases: a latent phase, an active phase, and a transition phase.

Second Stage of Labor

The second stage of labor lasts from full dilatation of the cervix to birth of the fetus.

The Third Stage of Labor

The third stage of labor lasts from the birth of the fetus until the placenta is delivered.

The Fourth Stage of Labor

The fourth stage of labor lasts about two hours after delivery of the placenta. It is the period of immediate recovery, when homeostasis is reestablished.

Terms Related to Labor

1. **Dystocia**: difficult labor due to impaired uterine forces, an abnormal position (presentation) of the fetus, inadequate birth canal
2. **Puerperium**: the period when the reproductive organs and maternal physiology returns to the prepregnancy state. It usually last for about six weeks.
3. **Involution**: reduction in the size of the uterus
4. **Lochia**: uterine discharge of blood and serous fluid that usually last for two to four weeks after delivery

28.6 Adjustments of the Infant at Birth

The Neonatal Period

- The four-week period immediately after birth is referred to as the neonatal period.
- The physical status of the infant is assessed one to five minutes after birth based on five signs: heart rate, respiration, color, muscle tone, and reflexes (tested by slaps on the feet).
- Each observation is given a score of 0 to 2 (**Apgar score**).
- An Apgar score of 8–10 indicates a healthy baby.

First Breath

- First breath is triggered as the placenta fails to remove carbon dioxide.
- Accumulation of carbon dioxide in the blood causes acidosis.
- Acidosis causes the respiratory control centers in the brain to trigger the first respiration.
- The first breath is painful due to the tiny airways and the collapsed lungs.
- Umbilical arteries and vein constrict and become fibrosed.
- The ductus venosus collapses and is converted to the ligamentum venosum, connecting the aorta and the pulmonary trunk.
- The foramen ovale is closed by two flaps of heart tissue.
- The remnant of the foramen ovale is the fossa ovalis.

The Transitional Period

- At birth, the infant's pulse may be from 120 to 160 per minute.
- The infant's oxygen demand increases, which stimulates an increase in erythrocyte and hemoglobin production.
- The white blood cells count at birth is very high, up to 45,000 cells per cubic millimeter, and decreases rapidly by the seventh day.
- Temporary jaundice may occur due to the infant's inability to control the production of bile pigment.

Lactation

- Lactation is the production of milk by the mammary gland.
- Prolactin (PRL) is the principal hormone promoting lactation.
- Prolactin levels increase as pregnancy progresses because estrogen and progesterone inhibit
    PRL from being effective.
- Following delivery, estrogen and progesterone decrease, and lactation begins.

Positive Feedback Mechanism of the Milk Let-down Reflex

- Stimulation of pressoreceptors in nipples by suckling infant sends efferent impulses to the hypothalamus.
- The hypothalamus sends efferent impulses to the posterior pituitary, where oxytocin is stored.
- Oxytocin is released from the posterior pituitary gland and stimulates myoepithelial cells of the breast to contract.
- Alveolar glands respond by releasing milk through the ducts of the nipples.

28.7 Genetics

28.7.1 The Vocabulary of Genetics

1. **Somatic cells:** Human cells, except the gametes, contain a diploid number of chromosomes (46) that is made up of 23 pairs of homologous chromosomes. Two sets are sex chromosomes (XY = male and XX = female). The other 44 are the 22 pairs of autosomes and guide the expression of other traits.
2. **Karyotype:** diploid chromosomal complement displayed in homologous pair
3. **Genome:** genetic (DNA) makeup containing two sets of genetic instructions from parents
4. **Gene pairs (alleles):** matched genes at the same locus (location) on homologous chromosomes
5. **Homozygous:** Two alleles controlling a single trait are the same.
6. **Heterozygous:** Two alleles controlling a trait are different.
7. **Dominant:** One allele masks or suppresses the expression of its partner.
8. **Recessive:** The allele that is masked or suppressed.

Genotype and Phenotype

- Genotype: individual's genetic makeup (homozygous or heterozygous) for the various types of gene pairs
- Phenotype: the way the genotype is expressed in an individual

28.7.2 Sexual Sources of Genetic Variation

- **Segregation and independent assortment of chromosomes:** The members of the allele pair determining each trait are distributed to different gametes during meiosis, and alleles on different pairs of homologous chromosomes are distributed independently of each other. The number of different gametes resulting from independent assortment of homologous pairs during meiosis I is 2^n, where n= the number of homologous pairs.
- **Crossover of homologues and gene recombination:** Paternal and maternal chromosomes exchange gene segments, giving rise to recombinant chromosomes with

mixed contributions from each parent.

- **Random Fertilization**: Gametes produced during gametogenesis contain various variations as a result of independent assortment and random crossovers. Also, a single egg will be fertilized.

Incomplete Dominance (Intermediate Inheritance)
- The heterozygote has a phenotype intermediate between those of homozygous dominant and homozygous recessive individuals. An example in humans is the sickle cell anemia trait. It causes a substitution of one amino acid in the β chain of hemoglobin.

Multiple-allele Inheritance
- Some genes exhibit more than two alternate forms. An example is the ABO blood types Three alleles determine the ABO blood types in humans: IA, IB, and I. Each of us receives two of these. IA and IB alleles are codominant, and both are expressed together when present. The I allele is recessive to the other two alleles.

Sex-linked Inheritance
- Inherited traits are determined by genes on the sex chromosomes. A gene found only on the X chromosome is said to be X-linked (e.g., hemophilia, color blindness).

Polygene Inheritance
- Many inherited traits do not follow the typical Mendelian genetics; instead, many phenotypes depend on several different gene pairs at different locations acting in tandem. For example, anAABBCC genotype results in a dark-skinned individual while aabbcc confers fair skin. Other examples are height and eye color.

Review Questions

1. Fertilization of the ovum usually occur in the _____ of the fallopian tube.
2. The endometrium is digested by enzymes released from the _____.
3. Sperm cannot fertilize an egg until they undergo _____.
4. What happens during amphimixis?

_____.

5. _____ are identical cells that are produced by early cleavage.
6. The embryo is formed by the

_____.

7. The germ layers are formed during the process of _____.
8. Neural tissues are formed by the _____ layer.
9. Muscle is formed by the _____ layer.
10. The urinary bladder is formed by the _____ layer.
11. The extraembryonic membrane that forms blood is the _____.
12. _____ is the time spent in prenatal development.
13. The outer layer of blastocyst cells that provides nourishment for the embryo is the _____.
14. The hollow cavity within the blastocyst is the _____.
15. The _____ is formed by the allantois, blood vessels, and yolk sac.
16. The blood vessels that carry blood to the placenta are the _____.
17. The blood vessels that carry blood away from the placenta are

_____.

18. The period from one month to two years is known as _____.
19. An individual's entire genetic makeup is called his or her _____.
20. The genes that are expressed in an individual produce the _____.
21. A(n) _____ allele will always be expressed regardless of what the other allele happens to be.
22. Genes that appear on the X chromosomes are said to be _____.
23. Arrange the following terms into their correct time sequence:
    1) blastocyst
    2) fertilization
    3) implantation
    4) zygote
    5) development of chorionic villi
    6) cleavage
24. Give an example of multiple-allele inheritance.

    _____

25. Name two traits that are determined by polygene inheritance.

    _____

26. List the four stages of labor.

    _____

27. The term parturition means

    _____.

28. List the three hormones that work cooperatively to stimulate the maturation of the breast for lactation.

    _____

    _____

29. What is the function of human chorionic gonadotropin hormone?

    _____

    _____

    _____

30. List two events that prevent polyspermy in humans during fertilization.

    _____

    _____

    _____

ANSWER KEY TO REVIEW QUESTIONS

Chapter 1

1. physiology
2. homeostasis
3. receptor; effector
4. Autoregulation
5. Extrinsic regulation
6. feedback
7. disease
8. prone
9. parasagittal
10. elbow
11. on the heart itself
12. pericardial sac; pericardial cavity
13. thoracic, abdominopelvic
14. peritoneum
15. urinary
16. distal
17. superior
18. embryology
19. integumentary system
20. lymphatic

Chapter 2

1. nucleus
2. energy levels
3. compound
4. cations
5. anions
6. exergonic
7. endergonic
8. Enzymes
9. Organic
10. solution
11. Electrolytes
12. hydrophilic
13. 18
14. Organic molecules
15. covalent
16. ionic
17. ATP
18. Pentose sugar; phosphate; a nitrogen base
19. nucleotides
20. nature of the R group
21. amino acids
22. triglycerides`

23. carbohydrate
24. isomers
25. synthesis
26. the number and arrangement of its electrons.
27. neutrons in the nucleus
28. protons in the atom
29. nuclei
30. atoms

Chapter 3

1. cytosol; organelles
2. inclusions
3. Somatic
4. ligands
5. resting membrane potential
6. mitosis
7. Replication
8. differentiation
9. storage and synthesis of molecules
10. codon
11. phagocytosis
12. endocytosis; exocytosis
13. crenation; lysis
14. apoptosis
15. hypertonic
16. smooth endoplasmic reticulum
17. active transport
18. hypotonic
19. Movement of materials from an area of high concentration to an area of low concentration
20. sodium; potassium

Chapter 4

1. mesothelium
2. endothelium
3. Ground substances
4. matrix
   plasma
6. membranes
7. lymph
8. cardiac muscle
9. epithelial; connective; muscular; neural tissues
10. Tendons; ligaments
11. Connective tissue proper; fluid connective tissue; supporting connective tissue
12. plasma cells
13. endocrine
14. merocrine
15. smooth

16. parietal peritoneum
17. mucous membranes
18. hyaline
19. bone
20. dense irregular connective tissue
21. collagen
22. adipocytes
23. lining the ducts of sweat gland
24. thyroid gland
25 basement membrane

Chapter 5

1. nails; sweat glands; sebaceous glands; hair follicles
2. epidermis; dermis
3. keratinocytes
4. stratum germinativum (basale)
5. stratum lucidum
6. stratum germinativum
7. papillary
8. reticular
9. sebaceous
10. apocrine

11. merocrine
12. It forms "goose bumps"
13. When the skin is stretched beyond its elastic limit
14. Abnormally large number of collagen fibers and relatively few blood vessels at the repair site
15. The thinning of the epidermis and decline of elastin
16. elastin
17. Increased melanin production
18. a bluish skin coloration
19. On an infant before it is born
20. seborrheic dermatitis
21. albinism
22. vitiligo
23. carcinoma
24. melanoma
25. hypothermia
26. severe pain, generalized swelling, and edema
27. stratum corneum
28. dermis
29. dilation of capillaries in the dermis
30. melanin

Chapter 6

1. osteocyte.
2. osteoblasts

3. osteoclasts
4. hydroxyapatite
5. diaphysis
6. epiphysis
7. metaphysis
8. Articular cartilage
9. ossification
10. calcification
11. process
12. Trochlea
13. Foramen
14. 206
15. osteogenesis
16. trabeculae
17. simple
18. compound
19. greenstick
20. Colle's
21. Osteoprogenitor
22. Yellow bone marrow
23. Osteomalacia
24. absent
25. increased
26. mesenchymal
27. medullary cavity

28. osteon
29. cartilage mode30. endosteum
31. blood vessels

Chapter 7

1. Occipital
2. Sagittal
3. Frontal, parietal, and occipital
4. Temporal and zygomatic
5. Temporal
6. Tear glands
7. Mastoid process
8. Styloid process
9. Stylomastoid foramen
10. Mandibular fossa
11. Pituitary
12. Crista galli
13. Scoliosis
14. Axis
15. Transverse process
16. Temporal bone
17. Support the larynx

18. It helps to hold the head in the upright position
19. Occipital condyles
20. Corona
21. Frontalis
22. Kyphosis
23. Lordosis
24. Scoliosis
25. Vertebral arteries and veins
26. Premature ossification of the fontanels occurring in infancy or early childhood
27. Skull, auditory ossicles, hyoid bone, vertebral column, and thoracic cage
28. Lightens skull and resonance
29. Sphenoid
30. Mandibular foramen

Chapter 7
Skeletal System: Appendicular

1. Scapula
2. Femur
3. Fibula
4. Humerus
5. Radius
6. Coxal
7. Clavicle
8. Tibia
9. Colle's
10. pelvic outlet
11. phalanges
12. metacarpals
13. acetabulum
14. glenoid fossa
15. patella
16. heel bone
17. scapula
18. pubic symphysis
19. pelvic brim
20. pelvic inlet (cavity)

Chapter 8
Articulation

1. synarthrosis
2. amphiarthrosis
3. synarthrosis
4. synostosis
5. syndesmosis
6. adduction
7. opposition
8. inversion

9. hinge
10. saddle
11. gliding
12. Coracoacromial
13. Gliding
14. gomphosis
15. hyperextension
16. abduction
17. flexion and extension
18. elevation
19. synchondrosis
20. bursae
21. protraction
22. retraction
23. depression
24. sacroiliac
25. outward
26. nucleus pulposus
27. abduction
28. bone to bone
29. muscle to bone
30. rheumatism

Chapter 9
Muscle Tissue

1. epimysium
2. perimysium
3. endomysium
4. tendon
5. the ability to produce large amounts of enzymes and structural proteins needed for contraction
6. myoblasts
7. sarcolemma
8. sarcoplasm
9. a transverse tubule and terminal cisternae
10. troponin molecules
11. neuromuscular junction
12. synaptic cleft
13. motor end plate
14. transverse tubules
15. Wave summation occurs when a muscle is stimulated repeatedly for several seconds with a constant stimulus and the amount of tension gradually increases to a maximum.
16. The phenomenon whereby a muscle produces peak tension with rapid cycles of contraction and relaxation
17. wave summation
18. complete tetanus
19. recruitment

20. isometric
21. aerobic respiration
22. It acts as an energy reserve in muscle tissue.
23. fatigue
24. Additional oxygen is required to metabolize the lactic acid produced during exercise
25. They have a low concentration of myoglobin.
    They produce powerful contractions.
26. slow
27. In the mitochondria
28. more permeable to sodium ions
29. It would cause spastic paralysis.
30. The event that occurs after death when muscle fiber runs out of ATP and calcium begins to leak from the sarcoplasm.

Chapter 10
Muscular System: Gross Anatomy

1. circular muscles
2. flex his thigh
3. pennate
4. tongue
5. chest
6. brbicularis oris
7. galea aponeurotica
8. superior nuchal line
9. move the external ear
10. skin of the chin
11. It wrinkles the brow
12. temporalis
13. palatoglossus
14. supraspinatus, infraspinatus, teres minor, subscapular ulnaris
15. fascicles

16. origin
17. insertion
18. inferior rectus
19. superior rectus
20. masseter
21. move the bone away from midline
22. move the bone toward the midline
23. increase the angle of a joint
24. decrease the angle of a joint
25. turn the hand palm posteriorly
26. turn the hand palm anteriorly
27. (Neck) sternocleidomastoid
28. (back) trapezius
29. (chest) pectorialis major

30. (abdominal wall) external oblique
31. (Shoulder) deltoid
32. (Upper arm) biceps brachii
33. (Buttocks) gluteus maximus
34. (Thigh) llastus lateralis)
35 (forearm) none
36. (legs) gastrocnemius

Chapter 11
Nervous Tissue

1. central
2. somatic
3. afferent
4. astrocytes
5. oligodendrocytes
6. microglia
7. Schwann cells
8. satellite cells
9. perikaryon
10. Nissl bodies
11. hillock
12. collaterals
13. telodendria
14. synaptic knobs
15. bipolar
16. unipolar
17. intracellular Na ions; extracellular K ions
18. passive
19. depolarization
20. 1. A graded depolarization brings an area of excitable membrane to threshold.
    2. Sodium channel activation occurs.
    3. Sodium ions enter the cell, and depolarization occurs.
    4. Sodium channels are inactivated.
    5. Voltage-regulated K channels open, and K moves out of the cell, inactivating polarization.
    6. Sodium channels regain their normal properties.
    7. A temporary hyperpolarization occurs.
21. All stimuli great enough to bring the membrane to threshold will produce identical action potentials
22. Type A fiber
23. norepinephrine
24. acetylcholine
25. peripheral
26. afferent (sensory)
27. Myelin
28. nodes

29. threshold
30. absolute refractory period
31. when sodium channels are opened
32. It results in local hyperpolarization.
33. a second EPSP arrives at a single synapse before the effects of the first have disappeared
34. several stimuli arrive at different locations on the same cell to produce action potentials (gated)
35. Gated

Chapter 12
The Spinal Cord, Spinal Nerves, and Spinal Reflex

1. columns
2. cervical; lumbar enlargements
3. filum terminale
4. pia mater
5. horns
6. gray commissure
7. bundles of axons that share common origins, destinations, and functions (Ascending, descending, and transverse tracts)
8. epineurium
9. cell bodies of sensory neurons
10. axons of motor neurons
11. axon of a sensory neuron
12. white ramus communicantes
13. dorsal ramus
14. cervical
15. leg
16. ventral ramus
17. plexus
18. phrenic
19. brachial
20. Brachial
21. flexor
22. tendon
23. dermatome
24. nerve plexus
25. Somatic
26. Intersegmental reflexes
27. quadriplegia
28. paraplegia
29. phrenic nerve
30. Incoming sensory information would be disrupted.

Chapter 13
The Brain and Spinal Nerves

1. medulla oblongata
2. medulla oblongata
3. medulla and pons
4. cerebellum
5. pneumotaxic center
6. corpora quadrigemina
7. postural reflexes
8. emotional behavior
9. Fornix; mammillary bodies; septum pellucidum; amygdaloid nucleus
10. tuber cinereum
11. mammillar body
12. left and right thalami
13. hypothalamus
14. water balance
15. hypothalamus
16. parietal lobe
17. temporal lobe
18. precentral gyrus
19. occipital lobe
20. parietal lobe
21. temporal lobe
22. lateral fissure
23. motor cortex
24. frontal lobe
25. cerebral hemisphere
26. gyrus
27. association tracts
28. precentral gyrus
29. corpus callosum
30. longitudinal fissure
31. trochlear !V
32. trigemina, mandibular division
33. accessory XI
34. vestibulocochlear VIII
35. amygdala in association with the hypothalamus

Chapter 14
Sensory, Motor, and Integration

1. nociceptors
2. nociceptors
3. thermoreceptors
4. Ruffini corpuscle
5. proprioceptors
6. baroreceptors
7. fasciculus gracilis

8.   posterior spinocerebellar
9.   lateral spinocerebellar
10. anterior spinothalamic
11. second-order
12. third-order
13. Sensory neurons from each body region synapse in specific brain regions.
14. anterior corticospinal tract
15. lateral corticospinal tract
16. vestibulospinal
17. reticulospinal
18. tectospinal
19. Voluntary control over skeletal muscle
20. preganglionic
21. Each receptor has a characteristic sensitivity.
22. A generator potential is a membrane depolarization that leads to an action potential in an excitable sensory membrane.
23. Baroreceptors
24. Involuntary regulation of eye, head, neck and arm position in response to visual or auditory stimuli will be affected
25. tectum
26. homunculus
27. reticular activating system
28. memory engrams (nucleic acid)
29. stereognosis
30. C

Chapter 15
The Special Senses

1.   Production of new supporting cells
2.   Chemical
3.   olfactory bulb
4.   levator palpebrae superioris
5.   fovea centralis
6.   sclera
7.   sclera; cornea
8.   macula lutea
9.   ciliary processes
10. The optic disk
11. retina
12. Contraction of the ciliary muscle
13. Conversion of rhodopsin to lumirhodopsin
14. lumirhodopsin
15. rhodopsin
16. Retinene
17. occipital
18. Interprets color and depth perception

19. external auditory canal
20. Equalization of air pressure in the middle ear
21. Detects rotational acceleration or deceleration
22. To transmit and amplify sound waves
23. It analyzes the frequency of sound
24. cochlea
25. saccule; utricle; semicircular canals
26. Light rays come into focus behind the retina of the eye
27. endolymph
28. basilar membrane
29. cataract
30. trachoma

Chapter 16
Autonomic Nervous System

1. autonomic ganglions
2. unmyelinated
3. They carry fibers that synapse in collateral ganglia.
4. stomach, liver and pancreas
5. inferior mesenteric ganglion
6. cholinergic
7. open sodium channels when stimulated
8. acetylcholine
9. An organ receives innervation from both sympathetic and parasympathetic nerves
10. Memories that can be voluntarily retrieved and verbally expressed
11. Conversion of a short-term memory to a long-term memory
12. A state of consciousness characterized by difficulties with spatial orientation, memory, language, and changes in personality
13. sympathetic
14. short; long
15. parasympathetic
16. Visceral reflexes
17. mimetics
18. deep
19. cholinergic
20. adrenergic

Chapter 17
Endocrine System

1. It converts ATP to cyclic AMP.
2. Hormones whose primary site of action are other endocrine glands.

3. It is the final activator that causes the cell to carry out the functions for which it is genetically designed.
4. Hormone diffuses through the cell membrane. Hormone molecules bind to receptor sites. Receptor proteins are structurally changed. Hormone-receptor complex migrates to the nucleus. Genes become activated.
5. Receptors sites are stimulated. Adenylate cyclase is activated. ATP breakdown. Increase in cyclic AMP concentration.
6. cyclic AMP
7. melanocyte-stimulating hormone
8. acromegaly
9. release factors
10. luteinizing hormone
11. human growth hormone
12. Through a portal system
13. Addison's
14. Infundibulum
15. decreased amount of urine
16. lack of ADH
17. Uterus
18. kidney tubule
19. hypotyroidism
20. Hypofunctioning of the thyroid gland
21. hyperfunctioning of the thyroid gland
22. calcitonin; parathormone
23. zona glomerulosa
24. Rising potassium level in the adrenal cortex
25. Excessive amount of glucocorticoids
26. insulin
27. somatostatin (GHIH)
28. Decreases liver glycogen level
29. glucagon; epinephrine
30. Alcohol inhibits ADH secretion, leading to copious urine output.

Chapter 18
The Blood

1. reticulocytes
2. leukemia
3. Neutrophils
4. diapedesis
5. leukopenia
6. basophil; eosinophil; neutrophil
7. basophil
8. monocyte
9. phagocytosis

10. throbocyte
11. ADP
12. megakaryocytes
13. thrombin
14. Factor X
15. Plasmin
16. embolus
17. neither antibody a nor antibody b
18. neither agglutinogen a nor agglutinogen b
19. thrombus
20. chemotaxis

Chapter 19
The Heart

1. right atrium
2. left atrium
3. pectinate muscles
4. chordae tendineae
5. Epicardium
6. diastole
7. systemic circuit
8. pulmonary trunk
9. pulmonary trunk
10. coronary
11. SA node
12. SA node→AV node→AV bundle→bundle branches→Purkinje fibers
13. QRS complex
14. End-diastolic volume
15. AV valve closes
16. AV valves and semilunar valves close
17. stroke volume
18. The cardiac output is directly related to the venous return
19. trabecular carnae
20. bradycardia

Chapter 20
Blood Vessels and Circulation

1. tunica media
2. continuous capillaries
3. regulate blood flow through a capillary
4. If the level of carbon dioxide at the tissue increases
5. pulse
6. It causes angiotensinogen to be converted to angiotensin.
7. vertebra
8. internal jugular

9. axillary
10. circle of Willis
11. vasa vasorum
12. The process of cyclic changes in vessel diameter that occurs at the origin of a capillary
13. Collateral
14. Arteriole segment of a central channel
15. subclavian artery
16. common carotid artery
17. celiac artery
18. hepatic artery
19. superior mesenteric artery
20. gonadal arteries
21. renal arteries
22. internal jugular vein
23. phrenic vein
24. Lymphatic vessels
25. arteriovenous anastomosis
26. tunica media
27. decreases; increases
28. muscular
29. inferior vena cava
30. hepatic vein

Chapter 21
Lymphatic System

1. thoracic duct
2. Beneath the epithelial lining of the small intestine
3. white pulp
4. macrophages
5. When mast cells release histamine, serotonin, and heparin
6. fever
7. monocytes
8. When C1 binds to an antibody attached to an antigen
9. Immunity that is genetically determined and present at birth
10. active
11. plasma
12. cytotoxic T
13. cytotoxic T
14. Depress the responses of other T cells and B cells
15. Immunological surveillance
16. thymus
17. cytotoxic cell
18. lymphocytes; macrophages

19. helper T cell
20. When the cell is processing antigens
21. Langerhan cells
22. Kupffer cells
23. Opsonins
24. Sustains and enhances the alternative pathway of complement fixation
25. immunoglobins
26. The process by which the surface of a microorganism is covered with antibodies and complement, rendering it more likely to be phagocytized

Chapter 22
The Respiratory System

1. oropharynx

2. nasopharynx

3. lobules

4. compliance

5. hypoxia

6. anoxia

7. eupnea

8. tidal volume

9. expiratory reserve volume

10. inspiratory reserve volume

11. residual volume

12. alveolar collapse

13. When bicarbonate ions leave the red blood cells

14. Breathing that involves active inspiratory and expiratory movement

15. Inhibits the pneumotaxic and inspiratory centers

16. Gas volume is inversely proportional to pressure.

17. The resulting pain when pleural fluid is unable to prevent friction between the opposing pleural surfaces

18. Protects the lungs from damage due to overinflation

19. apneustic

20. carbon dioxide

Chapter 23
The Digestive System

1. lamina propria
2. Waves of muscular contractions that propels the content of the digestive tract from one point to another
3. pulp
4. vestibule
5. hydrochloric acid
6. pepsinogen
7. gastrin

8. rugae
9. gastric pits
10. Protein digestion
11. plicae and villi
12. ileum
13. cholcystokinin
14. gastric inhibitory peptide (GIP)
15. activate protein-digesting enzymes
16. common bile duct
17. lobule
18. entry of food into stomach
19. external pouches of the colon
20. three longitudinal bands of muscle located beneath the serosa of the colon
21. Amylase   enzyme that works on starch
22. Bile   emulsifies fat
23. lactase   enzyme that works on milk sugars
24. maltase   enzyme that converts maltose to glucose
25. pepsin   hydrochloric acid activates this enzyme
26. peptidase   enzyme from lining of small intestine that produces the end products
                     of amino acids
27. sucrase   enzyme that works on cane sugar
28. trypsin   pancreatic enzyme that works on protein
29. secretin
30. fats

Chapter 24
Metabolism

1. The process whereby a molecule of glucose is converted into two molecules of pyruvic acid,and two molecules of ATP are produced
2. ammonia
3. oxidative phosphorylation
4. chloride
5. zinc
6. Vitamin A   plays a role in maintaining epithelia and visual pigment synthesis
7. Vitamin D   required for proper bone growth and calcium absorption
8. Vitamin E   prevents the destruction of vitamin A and fatty acids
9. Vitamin K   essential for production of clotting factors
10. Thimine   acts as a coenzyme in decarboxylation reactions
11. Riboflavin   a constituent of the coenzyme FAD and FMN
12. Niacin   a constituent of coenzyme NAD
13. Panthothenic acid   a constituent of coenzyme A
14. glucogenesis
15. glycogenesis
16. Oxidative phosphorylation
17. mitochondria
18. Glycogenolysis occurs in the liver, Fat mobilization occurs. Gluconeogenesis occurs in the liver. Ketone bodies may be formed

19. remove hydrogen atoms from organic molecules and transfer them to coenzymes
20. absorptive

Chapter 25
The Urinary System

1. renal cortex
2. renal pyramids
3. Bundles of tissues that lie between pyramids and extend from the renal cortex toward the renal sinus.
4. renal pelvis
5. major calyces
6. Bowman's capsule; the glomerulus
7. distal convoluted tubule
8. juxtaglomerular apparatus
9. proximal convoluted tubule
10. absorption of ions, organic molecules, vitamins, and water
11. podocytes
12. loop of Henle
13. to produce a concentration gradient that will allow the nephron to produce a hypotonic filtrate
14. Less urine is produced.
15. Urine with a lower specific gravity is produced
16. papillary duct
17. juxtamedullary nephrons
18. vasa recta
19. The area of the urinary bladder bounded by the openings of the two ureters and the urethra
20. creatinine clearance

Chapter 26
Fluid, Electrolyte, and Acid–Base Balance

1. excrete less hydrogen ions and less sodium so bicarbonate ions are reabsorbed
2. excrete hydrogen ions in exchange of sodium ions
3. a weak acid and the salt of a weak acid
4. raise the pH
5. protein buffer
6. ANP
7. a weak acid
8. pH of arterial blood will fall below 7.41.
   Increased respiratory rate
   Lower concentration of bicarbonate ions in blood
   An increase in carbonic acid in the bloodstream
9. compensated acidosis
10. respiratory acidosis
11. Lightheadedness, agitation, dizziness, and tingling sensations

12. respiratory acidosis
13. respiratory acidosis
14. decreases
15. respiratory alkalosis
16. high
17. metabolic akalosis
18. aldosterone
19. (BHP + IFCOP) – (IFHP + BCOP) = NFP
20. decreases
21. Calcium   involved in blood coagulation and neurotransmitter release, muscle contraction
22. Phosphate   component of nucleic acids, some lipids and high-energy storage compounds
23. Potassium   intracellular ion important in establishing the polarity of the cell membrane
24. Sodium   extracellular ion important in establishing the polarity of the cell membrane
25. Chloride   the major extracellular ion important in establishing the pH of gastric juice
26. Magnesium   a major intracellular ion that activates enzymes involved in carbohydrate metabolism

Chapter 27
The Reproductive System
1. Regulates the amount of follicle stimulating hormone in circulation
2. perineum
3. Secrete testosterone and inhibin
4. primary spermatocytes
5. 23
6. four
7. controlling the rate of sperm production; forming the blood-testis barrier
8. epididymis
9. hypothalamus
10. vagina opening; anus
11. hymen
12. Progesterone
13. labia minora
14. endometrium
15. vagina; urethra; greater vestibular gland
16. fimbrae
17. labia majora
18. Stimulates corpus luteum to produce progesterone
19. estrogen and progesterone
20. spermatogonium
21. corpora cavernosa
22. ovulation
23. estrogen; progesterone

24. menarche

Chapter 28
Development and Inheritance

1. upper third (ampulla)
2. trophoblast
3. capacitation
4. The male and female pronuclei fuse
5. Blastomeres
6. inner cell mass of the blastocyst
7. gastrulation
8. ectoderm
9. mesoderm
10. endoderm
11. yolk sac
12. Gestation
13. trophoblast
14. blastocoele
15. umbilical cord
16. umbilical arteries
17. umbilical veins
18. infancy
19. genotype
20. phenotype
21. dominant
22. X-linked
23. 1) fertilization 2) zygote 3) cleavage 4) blastocyst 5) implantation 6) development of chrionic villi
24. ABO blood types in humans
25. Skin color and eye color
26. dilatation stage, expulsion stage, placental stage, and recovery stage
27. "bringing forth young"
28. Human placental lctogen (hPL), estrogens, and progesterone
29. It bypasses the pituitary-ovarian control to prompt the corpus luteum to continue to secrete progesterone and estrogen.
30. 1) depolarization of the membrane of the oocyte 2) release of calcium from the ER into the cytoplasm, which activates the oocyte to prepare for cell division

GLOSSARY

absolute refractory period: the interval during which a second action potential absolutely cannot be initiated, no matter how large a stimulus is applied

acetylcholinesterase: an enzyme that breaks down the neurotransmitter acetylcholine to acetic acid and choline at the synaptic cleft so the next nerve impulse can be transmitted across the synaptic gap

acrosome: a membrane-bound structure covering the cap of a spermatozoon containing hydrolytic enzymes for penetrating the oocyte

actin myofilament: thin myofilament within the sarcomere; composed of two F actin molecules, tropomyosin, and troponin molecules

action potential: a change in membrane potential in an excitable cell that acts as an electrical signal

activation energy: energy that must be overcome in order for a chemical reaction to occur or minimum energy that must be overcome in order for a chemical reaction to occur

active transport: transport of a substance across a cell membrane against the concentration gradient with the help of energy generated by the hydrolysis of an energy carrier such as ATP

adaptive immunity: adaptive defense that consists of antibodies and lymphocytes capable of differentiation of self from non-self and the tailoring of the response against a specific foreign invader

adenylate cyclase: an enzyme that synthesizes cyclic adenosine monophosphate or cyclic AMP from triphosphate (ATP). Cyclic AMP functions as a second messenger to relay extracellular signals to intracellular effectors.

agglutination: the clumping of aggregates of antigens or antigenic material with antibodies in solution

agglutinin: antibody that aggregates a particulate antigen (e.g., bacteria) and causes agglutination

agglutinogen: any of the antigens present on the outer surface of red blood cells

alkalosis: a condition in which the body fluids have excess base, defined as when blood pH is 7.45 or above.

allele: one of the alternative versions of a gene at a given location (locus) along a chromosome

amygdala: nucleus in the temporal lobe of the brain that forms part of the limbic system involved in emotion processing and memory

angiotensin I: formed by the action of renin on angiotensinogen

angiotensin II: active form of angiotensin; synthesized from angiotensin I; stimulates profound vasoconstriction and aldosterone secretion

B

B cells: type of lymphocytes, that when stimulated by a particular antigen, differentiate into plasma cells that react with the specific antigens

blood-brain barrier: a protective network of blood vessels and cells that filter blood flowing to the brain so that some substances, such as certain drugs, are prevented from entering brain tissues while other substances are allowed to enter freely

bone remodeling: a continuous process of bone resorption (destruction) and bone formation in response to hormonal and mechanical factors for the purpose of maintaining normal bone mass

C

calcitonin: a peptide hormone that acts via a specific receptor to inhibit osteoclast function, thereby promoting a decrease in calcium blood levels

carbonic anhydrase: a zinc-containing enzyme that catalyses a reversible reaction between carbon dioxide hydration and bicarbonate dehydration; facilitates the transfer of carbon dioxide from tissues to blood and from blood to alveolar sac

cardiac output (CO): amount of blood that is pumped by the heart per unit of time; measured in liters per minute (L/min)

cardiac reserve: difference between resting and maximum cardiac output; about 300 to 400 percent in a normal young adult

carotid body: a receptor rich in capillaries, at the spot the carotid artery branches in the neck, containing cells that sense the oxygen and carbon dioxide levels in the blood and from which messages are dispatched to the medulla to regulate the heart rate

cell-mediated immune response: an immune response that does not involve antibodies or complements but rather involves the activation of macrophages, natural killer cells (NK), antigen-specific cytotoxic T lymphocytes, and the release of various cytokines in response to an antigen

chemoreceptor: a receptor that is sensitive to chemical stimulation; located in the carotid and aortic bodies, taste buds, and olfactory cells in the nose

chloride shift: the exchange of chloride (Cl$^-$) and bicarbonate (HCO$_3$) between plasma and the erythrocytes occurring whenever HCO$_3^-$ is generated or decomposed within the erythrocytes

choroid plexus: a structure made from tufts of villi within the ventricular system; where cerebrospinal fluid (CSF) is produced by modified ependymal cells

chromatin: mass of genetic material composed of DNA and proteins (nucleosome) that condense to form chromosomes just before cell division

cloning: the process used to create an exact genetic copy of another cell, tissue, or organism

coagulation: the process by which the blood clots to form solid masses, or clots

commissure: a tract of nerve fibers passing from one side to the other of the spinal cord or brain

complement: group of serum proteins that move freely through the bloodstream that work with the immune system for the development of inflammation

D

depolarization: the reduction of a membrane potential to a less negative value as a result of the influx of cations like sodium and calcium

dermatome: an area of skin innervated by sensory fibers from a single spinal nerve

desmosome: the point of attachment between cells consisting of dense plate in each adjacent cell

desquamate: peeling off of the superficial cells of the outermost layer of the skin (stratum corneum)

diapedesis: the movement of white blood cells through intact capillary walls into surrounding body tissue

E

electron transport chain: series of electron carriers in the inner mitochondrial membrane that receive electrons from NADH and $FADH_2$, using the electrons in the formation of ATP and water

endocytosis: the process by which animal cells engulf particulate materials such as bacteria, cellular debris, and macromolecules

endosteum: the thin layer of cells lining the medullary cavity of a bone

epiphyseal plate: a thin layer of cartilage between the epiphyses

eponychium: outgrowth of the skin that covers the proximal and lateral borders of the nail

erythroblastosis fetalis: a severe hemolytic disease of a fetus or newborn infant caused by the production of maternal antibodies against the fetal red blood cells, usually involving Rh incompatibility between an Rh-negative mother and an Rh-positive fetus

erythropoiesis: the formation of erythrocytes in the bone marrow

erythropoietin: a hormone produced by the kidney that promotes the formation of red blood cells in the bone marrow

extrinsic clotting pathway: the mechanism that produces fibrin following tissue injury, beginning with formation of tissue thromboplastin found outside the blood

F

falciform ligament: fold of the peritoneum extending to the surface of the liver from the diaphragm and anterior abdominal wall

falx cerebelli: a fold of the dura mater separating the cerebellar hemispheres

falx cerebri: a fold of the dura mater in the longitudinal fissure separating the cerebral hemispheres

fibroblast: a cell that synthesizes extracellular matrix and collagen

fimbrae: numerous fingerlike processes on the distal part of the infundibulum of the uterine tube

fovea centralis: a depression toward the center of the retina containing only cone cells where the vision is most acute

G

ganglion: a biological tissue mass of nerve cell bodies

gap junctions: cylindrical channels between animal cells that allow small molecules and ions to pass from the inside of one cell to the inside of an adjacent cell

gastric pit: a depression in the stomach containing mucosa, parietal, and chief cells

gluconeogenesis: metabolic pathway that results in the generation of glucose from non-carbohydrate carbon substances such as lactate, glycerol, and glucogenic amino acids

glycogenesis: formation of glycogen from glucose

goblet cell: specialized epithelial cells found in the mucous membrane of the stomach, intestine, and respiratory passages that secrete mucus

H

hematocrit: the proportion of red blood that consists of packed red blood cells; expressed as a percentage by volume

hemostasis: stoppage of bleeding or hemorrhage

hippocampus: major component of human brain; belongs to the limbic system and functions in memory and spatial navigation

histamine: substance that plays a major role in allergic reaction; dilates blood vessels

homeostasis: ability of the body to physiologically regulate its inner environment to ensure its stability in response to fluctuations in the environment

hyperpolarization: a change in the value of resting membrane potential towards a more negative value

hypophyseal portal system: a system of blood vessels that link the hypothalamus and the anterior pituitary gland

I

immunity: resistance of an organism to infection or disease

inflammation: body's basic response to injury

inhibitory postsynaptic potential (IPSP): hyperpolarization in the membrane of a postsynaptic neuron caused by the binding of an inhibitory neurotransmitter from a presynaptic receptor; makes it difficult for a postsynaptic neuron to generate an action potential

intercalated discs: dark staining, transversely oriented bands that link cardiac muscle cells together

intramembranous ossification: the formation of bone within a connective tissue without the prior development of a cartilaginous model

intrinsic pathway: a sequence of reactions leading to fibrin formation, beginning with the contact activation of factor XII, and resulting in the activation of factor X to initiate the common pathway of coagulation

ipsilateral: a reflex response that affects the same side as the stimulus

J

juxtaglomerular apparatus: the cells near the glomerulus in the kidney that stimulate the secretion of the adrenal hormone, which plays a role in renal autoregulation

juxtaglomerular cells: specialized smooth muscle cells in the wall of the afferent arteriole in the kidney that secrete the enzyme renin

K

ketoacidosis: a reduction in the pH of body fluids due to the presence of large numbers of ketonebodies

killer T cells: T cells with CD8 receptors that recognize antigens on the surface of virus-infected cells, bind to the infected cells, and kill them

Kupffer cells: specialized macrophages located in the liver, lining the walls of the sinusoids, that form part of the reticuloendothelial system (RAS)

L

lamina propria: thin vascular layer of connective tissue beneath the epithelium of an organ
ligamentum arteriosus: the fibrous strand in an adult that is a remnant of the ductus arteriosus of the fetal stage
limbic system: a group of nuclei and centers in the cerebrum and diencephalon that are involved emotional states, memories, and behavioral drives
lymphokines: chemicals secreted by activated lymphocytes
lysosome: spherical organelle that contains digestive enzymes
lysozyme: an enzyme present in some exocrine secretions such as tears, saliva, and milk that has antibiotic properties

M

macula densa: area of closely packed specialized cells lining the wall of the distal tubule, adjacent to the glomerulus and the juxtaglomerular cells
major histocompatibility complex: a group of genes that code for cell surface histocompatibility antigens and that are principal determinants of tissue type and transplant compatibility
mediastinum: the central tissue mass that divides the thoracic cavity into two pleural cavities; contains the heart, esophagus, trachea, phrenic nerve, cardiac nerve, thoracic duct, thymus, and lymph nodes
medullary rhythmicity center: the center of the medulla oblongata that sets the background pace of respiration; includes inspiratory and expiratory centers
mesenchyme: embryonic or fetal connective tissue
metarteriole: a vessel that connects an arteriole to a venule and that provides blood to a capillary plexus
MHC proteins: surface antigen that is important to the recognition of foreign antigens and that plays a role in the coordination and activation of the immune response; also referred to as *human leukocyte antigen* (HLA)
mucosa-associated lymphoid tissue: the extensive collection of lymphoid tissues linked with the digestive system

N

neoplasm: abnormal new growth of tissue; tumor
neurohypophysis: the posterior (or neural) lobe of the pituitary gland, which stores oxytocin and vasopressin
neuromuscular junction: a synapse between a neuron and a muscle cell
nicotinic receptor: acetylcholine receptors on the surface of sympathetic and parasympathetic ganglion cells; responds to the compound nicotine
nucleus pulposus: a jelly-like substance in the central region of an intervertebral disc

O

opsonization: the rendering of bacteria and other cells subject to phagocytosis

optic chiasm: a point near the thalamus and hypothalamus at which portions of each optic nerve cross over

optic tract: the tract over which nerve impulse from the retina are transmitted between the optic chiasm and the thalamus

osmotic pressure: the pressure exerted by the flow of water through a semipermeable membrane separating two solutions with different concentrations

osteon: basic unit of structure of compact bone, comprising of a hervasian canal and its concentrically arranged lamellae

P

parathyroid hormone: a protein hormone released by the parathyroid gland; the most major regulator of the body's calcium and phosphate levels

perichondrium: the layer of connective tissue that surrounds a cartilage

perilymph: an extracellular fluid located within the cochlea, the scala tympani, and scala vestibule

peritoneal cavity: a potential space between the parietal peritoneum and visceral peritoneum

precentral gyrus: the primary motor cortex of a cerebral hemisphere, located anterior to the central sulcus

R

rami communicantes: bundles of nerve fibers connecting a sympathetic ganglion to spinal nerve, consisting of gray rami and white rami

repolarization: the movement of the transmembrane potential away from a positive value and toward the resting potential

respiration: the exchange of gases between cells and the environment; includes pulmonary ventilation, external respiration, internal respiration, and cellular respiration

reticular formation: a comprehensive network of nerves that is formed in the central area of the brain stem

retroperitoneal: anatomical space in the abdominal cavity behind the peritoneum

rugae: mucosal folds in the lining of an empty stomach that stretch when food enters

S

sarcomere: contractile unit of a myofibril, delimited by the Z bands, along the length of the myofibril

sarcoplasm: cytoplasm of a muscle fiber

sciatic nerve: the longest and widest single nerve in the human body; begins in the lower back and runs through the buttock and down the lower limb

sinoatrial (SA) node: the natural pacemaker of the heart; situated in the wall of the right atrium

sliding filament theory: explanation of how muscles produce force; explains that the thin and thick filament within the sarcomere slide past one another, shortening entire length of the sarcomere

stroma: the connective tissue framework of an organ

suppressor T cells: cells that express the CD8 transmembrane glycoprotein and that close down the immune response after invading organisms are destroyed

syncytium: a multinucleate mass of cytoplasm, produced by the fusion of cells or repeated mitoses without cytokinesis

synergist: a muscle or agent that acts with another

T

t cells: group of white blood cells known as lymphocytes that play a central role in cell-mediated immunity

tectospinal tract: a tract that conveys nerve impulses from the midbrain to the spinal cord in the cervical region

telodendria: structures at the terminal branches of axons that contain neurotransmitters

trabecula: a connective tissue partition that divides an organ

tract: a bundle of axons in the central nervous system

twitch: a single stimulus-contraction-relaxation cycle in a skeletal muscle

U

umbilical cord: the connecting cord, containing two arteries and one vein, from the developing embryo or fetus to the placenta

ureters: muscular, narrow tubes, lined by transitional epithelium, that carry urine from the kidney to the urinary bladder

urethra: a muscular tube that carries urine from the urinary bladder to the exterior

uterus: the muscular, hollow, pear-shaped organ of the female reproductive tract in which implantation, placenta formation, and fetal development occur

V

vagina: a muscular tube lined with mucus membranes, extending between the uterus and the vestibule

vasoconstriction: the narrowing of blood vessels by smooth muscles in the tunica media

vasodilation: the widening of blood vessels resulting from relaxation of the muscular wall of the blood vessels

vasomotion: changes in the pattern of blood flow through a capillary bed in response to changes in the local environment

W

white matter: one of the two components of the central nervous system made up of myelinated axons

Z

zygote: fertilized egg that contains a diploid number of chromosomes

References

1. Frederic H. Martini: *Fundamentals of Anatomy and Physiology, ed 4*, Upper Saddle River, N.J., 1989, Prentice Hall

2. Seely Stephens Tate: *Anatomy and Physiology, ed 8*, New York, 2008, McGraw-Hill

3. Gerald J. Tortora: *Principles of Anatomy and Physiology, ed 11*, Hobokin, N.J., 2005, Wiley

4. Elaine N. Marieb: *Human Anatomy and Physiology, ed 5*, N.Y., 1999, Benjamin Cummings

5. Bobak, Lowdermilk, and Jenson: *Maternity Nursing, ed 4*, N.Y., 1995, Mosby

6. Lucia Tranel and Alice Mills: *Instructor's Resource Guide*, N.J., 1995, Prentice Hall

7. Dunn: Professors Resource Manual to accompany Tortora/Grabowsskyi's *Principle of Anatomy and Physiology, ed 8*, N.Y., 1996, HarperCollinsCollegePublishers.

8. Janice Yoder Smith, *Concept Maps* To accompanyTortora/Grabowski *Principles of Anatomy and Physiology, ed 7*, HarperCollinsCollege Publishers.

Manufactured By: RR Donnelley
                     Breinigsville, PA USA
                     January, 2011

Windows 7 @ cuny

requser ab1

umesh@mec.cuny.edu